Innovation in Technology Alliance Networks

Innovation in Technology Alliance Networks

Charmianne E.A.V. Lemmens

Assistant Professor of Organization Science,
Eindhoven University of Technology (TU/e),
The Netherlands

Edward Elgar
Cheltenham, UK • Northhampton, MA, USA

Published by
Edward Elgar Publishing Limited
Glensanda House
Montpellier Parade
Cheltenham
Glos GL50 1UA
UK

Edward Elgar Publishing, Inc.
136 West Street
Suite 202
Northampton
Massachusetts 01060
USA

A catalogue record for this book
is available from the British Library

ISBN 1 84376 990 5

Printed and bound in Great Britain by MPG Books Ltd, Bodmin, Cornwall

Contents

List of Figures	*viii*
List of Tables	*ix*

1 Introduction to the Study	1
1.1 Introduction	1
1.2 The driving forces in alliance network formation	3
1.3 Alliance network formation from a social network perspective	4
1.4 The driving forces in alliance block formation	5
1.5 Alliance block formation from a social network perspective	7
1.6 Alliance block formation from a network evolutionary perspective: the socio-technical system	9
1.7 Research questions	10
1.8 Outline of the study	12

2 Definitions, Data and Methodology	14
2.1 Introduction	14
2.2 Definitions	14
2.3 Data sources	16
2.4 Methodology	17
2.5 Variables	21
2.6 The microelectronics industry	27

3 The Enabling Effect of Embeddedness	32
3.1 Introduction	32
3.2 Theoretical background	32
3.3 Propositions	33
3.4 Sample and data	37
3.5 Methodology	37
3.6 Results	38
3.7 Discussion and conclusion	40

4 The Constraining Effect of Embeddedness   42
  4.1 Introduction   42
  4.2 Theoretical background   43
  4.3 Propositions   44
  4.4 Sample and data   46
  4.5 Methodology   47
  4.6 Results   48
  4.7 Discussion and conclusion   50

5 Alliance Block Members: Who Are They?   52
  5.1 Introduction   52
  5.2 Theoretical perspectives on alliance blocks   52
  5.3 Strategic groups vs. alliance blocks   54
  5.4 Attributes of alliance block members   56
  5.5 Discussion and conclusion   59

6 The Innovative Performance of Block Members   63
  6.1 Introduction   63
  6.2 Theoretical background   64
  6.3 Hypotheses   66
  6.4 Sample and data   70
  6.5 Methodology   70
  6.6 Variables and measures   71
  6.7 Results   73
  6.8 Discussion and conclusion   77

7 Technological Change and Performance of Block Members   80
  7.1 Introduction   80
  7.2 Theoretical background   81
  7.3 Hypotheses   81
  7.4 Sample and data   84
  7.5 Methodology   84
  7.6 Variables and measures   85
  7.7 Results   88
  7.8 Discussion and conclusion   92

8 Discussion and Conclusions   97
  8.1 Introduction   97
  8.2 The dual role of social structure in the alliance
      network formation process   98
  8.3 Block membership and its effect on innovative
      performance   102

8.4 Block membership and its effect on innovative
    performance under technological change   106
8.5 Final remarks   110

9 Summary   112

Appendix I   *115*
Appendix II   *116*
Appendix III   *119*
Appendix IV   *120*
Appendix V   *121*
Appendix VI   *122*

References   *126*

Index   *137*

Figures

1.1	Alliance blocks in the microelectronics industry	7
1.2	Conceptualization of research questions 2 and 3	12

2.1	Line connectivity in 1975–77	24
2.2	Line connectivity in 1976–78	25
2.3	Line connectivity in 1985–87	26
2.4	The number of alliances formed in microelectronics	30

3.1	Alliance block members and network size in 1970–2000	39
3.2	Number of alliance blocks and network size in 1970–2000	40

4.1	The enabling and constraining effects of embeddedness	46
4.2	Number of alliance block members in 1970–2000	49
4.3	Number of alliance blocks in 1970–2000	50

6.1	Over-embeddedness	69
6.2	The expected linear model 1	70
6.3	The expected curvilinear relation model 2	71
6.4	The curvilinear relation between innovative performance and block membership model 2	76

7.1	The expected interaction effect model 3	85
7.2	The curvilinear relation between innovative performance and the interaction effect model 3	90
7.3	The curvilinear relation model 3	93

Tables

2.1   Overview of variables used in the analysis   20

3.1   Ratio measuring in-/out-group strength   39

4.1   The alliance network formation process   42

5.1   One-sample t-test grouping variable   57
5.2   Results of discriminant analysis   58
5.3   Canonical discriminant function   59
5.4   Classification results   59

6.1   Pearson correlation coefficients of models 1 and 2   74
6.2   Regression estimates of models 1 and 2   75
6.3   Change statistics from linear to curvilinear model   77

7.1   Dissimilarity matrix   86
7.2   Variable technological change   86
7.3   Z-score normalizations   87
7.4   Change statistics interaction effect   89
7.5   Pearson correlation coefficients model 3   91
7.6   Regression estimates interaction effect model 3   92

1. Introduction to the Study

1.1 INTRODUCTION

The goal of this study is to build conceptually on the academic work of social structure and its effect on alliance formation patterns in social networks (for example Gulati, 1995a, 1998; Granovetter, 1992; Gulati and Gargiulo, 1999; Walker et al., 1997) by integrating these streams of research. Furthermore we aim empirically to study network evolution in inter-organizational networks in high-tech industries from a longitudinal perspective. More particular we aim at further theoretical development of the concept of alliance blocks[1] or 'technology driven constellations' (Gomes-Casseres, 1996; Das and Teng, 2002) by adopting a social network perspective (for example, Nohria, 1992; Gulati, 1998). Therefore we intend empirically to identify the social mechanisms that cause enabling and constraining effects of embeddedness in technology alliance blocks. The dynamics of inter-alliance networks are increasingly driven by these social mechanisms that follow from embeddedness and investing in social capital (Gulati, 1995a; Gulati and Gargiulo, 1999). These social mechanisms cause inter-organizational networks to self-generate, self-transform and self-reinforce in alliance blocks. Thus in this study we try empirically to identify the social component that is gaining importance over the technological aspect as a driving force in the network evolutionary process and in the formation of alliance blocks particularly, as firms increasingly look for trustworthy and preferential relations through replication of their existing ties to improve their innovative performance.

Moreover since it is a debated issue in the academic literature, this study intends to contribute conceptually as well as empirically to the current body of literature on social embeddedness and network-positioning strategies of firms in alliance networks – in alliance blocks in particular – and their effect on innovative performance (for example Rowley et al., 2000; Gulati et al., 2000; Gargiulo and Benassi, 2000; Coleman, 1988; Burt, 1992) under changing technological conditions (Madhavan et al., 1998; Bower and Christensen, 1995). In spite of a growing body of theoretical contributions, the academic literature and empirical research is rather inconclusive about the performance effects of group membership, which is the strongest form of social embeddedness. Given the emphasis on 'technology' alliances in this book, their effect will not be related to economic performance in general but to the innovative performance of companies (Hagedoorn and Duysters, 2002). Strategic technology alliances, through which companies acquire R&D-related knowledge, are expected to help them differentiate their innovative performance from other companies (Hagedoorn and Duysters, 2002). Hagedoorn and Schakenraad (1994) found a positive relation between technology-based alliances and their innovation rates.[2] Concerning the usefulness of engagement in technology alliances to improve innovative performance, Duysters and Hagedoorn (2000) found that strategic technology alliances should be used as a vehicle for developing core competences related to innovation in order to complement capabilities in the long run. Hence they can be used as monitoring devices to scan the most promising technologies. Because of the globalization of markets, the increasing complexity of technologies and rapid technological change and the increasing costs of R&D, technology alliances enable firms to explore several technological developments as well as exploit the most promising ones internally at the same time (Duysters and Hagedoorn, 2000).

Our empirical contribution manifests itself in that we try to establish the performance effects of firms embedded in alliance networks, which have multilateral collaborative technology agreements. We particularly focus on the performance effects of alliance block membership. To date, empirical evidence is lacking in that field. So far empirical studies have increasingly shown the effects of bilateral collaborative agreements on innovative performance (for example Das and Teng, 2003). Specifically, a study performed for the Dutch Ministry of Economic Affairs has empirically shown that bilateral collaborative agreements have positive effects on innovative performance for the firms involved. The study concluded amongst other things that a high level of alliance capability, similar knowledge backgrounds and a higher intensity of relationships increase innovativeness (De Man and Duysters, 2003). In our study we intend to address ourselves to the debated issue of the performance effects of firms connected through multilateral collaborative technology agreements and alliance blocks in particular. Therefore we empirically and conceptually build on the academic work in this field to show the effects of embeddedness on innovative performance (for example Rowley et al., 2000; Gulati et al., 2000; Gargiulo and Benassi, 2000; Coleman, 1988; Burt, 1992). By empirically building on Coleman's argument on closure advantages and Burt's argument on structural hole advantages regarding network position, we intend to establish the performance effects of alliance block membership. Hereby we investigate network positions that can involve pursuing either alliance block membership or a non-block membership strategy.

Concerning the moderating effect of technological change on network positioning strategies and innovative performance, we contribute empirically and conceptually to the academic work in this field (for example Madhavan et al., 1998) to show that the degree of uncertainty in the environment and required rate of innovation (Bower and Christensen, 1995) influence the appropriate network configurations.

1.2 THE DRIVING FORCES IN ALLIANCE NETWORK FORMATION

Interdependence and complementarities have been addressed as the most common explanation for firms forming inter-organizational ties (Richardson, 1972; Pfeffer and Nowak, 1976; Nohria and Garcia-Pont, 1991). These resource dependency perspectives (Pfeffer and Salancik, 1978) posit that external sources scarcity is the most important reason for engaging in collaborative agreements (Park et al., 2002). This stream of research has made significant progress in examining the factors – that is the exogenous dynamics – that determine the propensity of firms to form alliances. Specific motives for the formation of technology-based alliances have been raised, such as decreasing uncertainty and costs in R&D, technology transfer, technological leapfrogging, shortening of product life cycle, reducing the time-to-market, and motives related to market access and search for opportunities (Hagedoorn, 1993).

Recently however, the strategic alliance literature has made progress in advancing our understanding of the inter-alliance dynamics, that is how social factors, social relations and competitive tension among alliances affect the intent of creating, building and sustaining collaborative advantage through alliance formation (for example Gulati, 1995a, 1998; Walker et al., 1997; Gulati and Gargiulo, 1999; Chung et al., 2000). This so-called endogenous dynamic refers to with whom specifically alliances are formed

(Gulati, 1995a; Gulati and Gargiulo, 1999), as firms have several suitable partners at their disposal. In this context alliance formation is based on building preferential relationships characterized by trust, stability and rich exchange of information between partners (Dore, 1983; Powell, 1990; Gulati and Gargiulo, 1999). Most of these theoretical contributions on network evolution (see for example Gulati, 1998; Walker et al., 1997) assert that network formation proceeds through the formation of new relationships, building on the experience with existing firm ties. This stream of research thus has focused on the role of social structural context as an important driving factor in the alliance formation process (for example Gulati, 1995a; Walker et al., 1997; Gulati and Gargiulo, 1999; Chung et al., 2000). This social structural context refers to the fact that firms are embedded in a network of relations and have access to several qualified and resource-complementary partners, which influences their decision on whom to tie up with.

1.3   ALLIANCE NETWORK FORMATION FROM A SOCIAL NETWORK PERSPECTIVE

The social network perspective we adopt, addresses this social structural context driving the alliance formation process. It explains the collaborative behaviour of actors in terms of their position in networks of relationships (for example Nohria, 1992; Gulati, 1998). The perspective thus posits that actors are embedded in a network of social relations – their alliances. Embeddedness refers to the structure of a network of social relations that can affect the firm's economic action, outcomes and behaviour and that of its partners to whom it is directly or indirectly linked (for example Granovetter, 1992; Gulati, 1998). Thus embeddedness influences the firms' tying behaviour, because it enables preferential relations to emerge from the direct and indirect contacts firms have built up in their previous partnerships. By investing in these social relations through the replication of their existing ties, firms build up social capital (Burt, 1992). Social capital captures the shared values, norms and trust between alliance partners and is thus by its very nature dependent on history (Chung et al., 2000). Social capital thus enables firms to rely on both direct and indirect alliance experiences in partner selection (Chung et al., 2000) and hence allows them to shortcut the partner-selection process. In this way social capital generates returns as it enables firms to access and capture the embedded resources in their social relations (Lin,

1999). Moreover firms in network positions with higher social capital are likely to increase their inter-company relationships (Walker et al., 1997). From this we can conclude that social capital drives the network to self-organize, self-transform and self-reinforce, as social capital forms the basis upon which the actors establish future social relations (Gulati, 1998; Walker et al., 1997; Chung et al., 2000). In this way the network becomes a growing repository of information on the availability, reputation, competencies and reliability of prospective partners (Walker et al., 1997; Gulati, 1995a; Powell et al., 1996).

1.4 THE DRIVING FORCES IN ALLIANCE BLOCK FORMATION

Apart from engaging in these collaborative agreements to foster innovative renewal, firms increasingly adopt multiple collaborative arrangements for competitive gains (Guidice et al., 2003). In the alliance network formation process, technological positioning in the network depends very much on the competitive forces that shape the industry. Globalization of competition and the deepening industry convergence force firms to engage in global-scale production and acknowledge the cross-linking of industries through new technologies (Gomes-Casseres, 1996). Especially in high-tech sectors where technology positioning is crucial to firms' survival chances, 'competition through cooperation' (Gomes-Casseres, 1994, 1996; Doz and Hamel, 1998) has become a cornerstone of the firm's competitive strategy.

By establishing multiple collaborative agreements, firms tend to compete intensely with each other in several areas they are active in, resulting in 'co-opetition' behaviour (Gnyawali and Madhavan, 2001). Thus a firm's alliances can be instruments to withstand competition – by making enemies partners but can also impose stronger competition on others, as winning the alliance race entails access to better partners, resources or patents (Silverman and Baum, 2002). As these cooperative technology agreements among competitors proliferate (Gomes-Casseres, 1996; Gnyawali and Madhavan, 2001) technology competition becomes indispensable in the technology positioning strategy of the firms involved.

This actual explosion of collaborative agreements has led to a new form of competition: group-versus-group rather than company versus company (Gomes-Casseres, 1996; Guidice et al., 2003). The driving forces behind the formation of these technology-driven constellations are typically related to technology competition. Technology competition takes the form of multiple partner firms linked with each other through strategic alliances in groups or constellations (Das and Teng, 2002) 'competing against other such groups and against traditional single firms' (Gomes-Casseres, 1996: 3). Through multiple R&D collaboration in alliance blocks, innovators can capture the full benefit of their innovative activity through spillovers and externalities, as they now are able to share the costs and revenues of R&D projects, which can serve as an incentive to conduct further R&D (Sakakibara, 2002). Other important driving forces that incur group formation involve establishing industry standards as a result of standard battles among firms and entail (re)positioning strategies of companies (Gomes-Casseres, 1996; Das and Teng, 2002). A common theme behind these motivations is taking advantage of economies of scale and scope (Gomes-Casseres, 1996).

Research by Gomes-Casseres (1996) and by Doz and Hamel (1998) is among the first to have explored the increasing frequency of collaboration as a reflection of a fundamental shift from the traditional form of competition (firm vs. firm) to a new form (group vs. group).

As global competition continues to intensify and the number of technology-based alliances is increasing (Das and Teng, 2002), a more thorough understanding of this new form of competition – alliance-based technology competition through constellations (Figure 1.1) – is required to improve our understanding of how alliance networks evolve over time (Gomes-Casseres, 1996). Then, in these high-tech industries characterized by alliance races to get access to high-quality partners and their R&D capabilities (Gulati, 1995a), scarcity in the amount of available partners implies that firms have to move quickly to foreclose the competitors' partnering opportunities (Gomes-Casseres, 1994; Silverman and Baum, 2002). This behaviour can arouse an alliance network dynamic and explain the formation of these alliance blocks. Such group-versus-group competition does not however decrease the importance of the competition that takes place at the firm level, but it certainly alters the nature of competition as it increases the significance of a firm's alliances (Silverman and Baum, 2002). However the specific issue of group-based competition is beyond the scope of this research, since we focus on the formation of these alliance blocks and their performance effects.

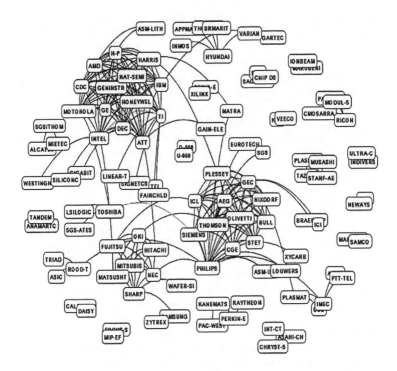

Source: Centre for Global Corporate Positioning (CGCP).

Figure 1.1 Alliance blocks in the microelectronics industry

1.5 ALLIANCE BLOCK FORMATION FROM A SOCIAL NETWORK PERSPECTIVE

In the network evolutionary process, and in the subgroup formation process in particular, both the social driving forces and technological drivers are important. Finding the right partners with complementary resource configurations is costly and time-consuming. As a consequence, firms tend to engage in local searches for forming their subsequent ties, based on the social capital (Burt, 1992) they have built up in their past partnerships. Engaging in preferential partnering thus tends to reduce the search costs of finding the right partners with complementary resource configurations, and eases the risk of opportunistic behaviour between the partners involved (Gulati and Gargiulo, 1999). Thus actors search for partners they trust and with whom they have had prior linkages. When prior linkages determine the formation of future linkages, those social relations are characterized as path-dependent (for example Gulati, 1995a; Levinthal and Fichman, 1988; Walker et al., 1997; Tsai, 2000). The local focus in search strategies in high-tech sectors is addressed by several organizational theories. Structural inertia theory (Hannan and Freeman, 1984), evolutionary economic theory (Nelson and Winter, 1982) and organizational learning theory (Cohen and Levinthal, 1990) all conclude that firms behave according to specific routines that have become institutionalized over time.

As firms build on these preferential relations (Dore, 1983; Powell, 1990; Gulati and Gargiulo, 1999) through replication of their ties, they become embedded in dense networks of relations. Engaging in new collaborations based on social capital through the replication of existing ties, typically results in the formation of densely connected cliques or blocks of collaborative relationships, which consist of firms that are all mutually connected through multiple alliances. These closely connected parts of the network are characterized by shared values, norms and trust between alliance partners. Such an environment provides a strong basis of trust and intimacy for the companies involved (Krackhardt, 1992; Brass et al., 1998; Granovetter, 1973) and hence provides the basis for further reproduction of this collective asset. Here a social dimension is apparent in the role of trust and shared value inducing reciprocity and knowledge sharing, as well as in preventing firms acting unethically (Brass et al., 1998; Kash and Rycoft, 2000).

The growth of alliance blocks can be limited by internal organizational factors, which can reduce the net benefits of adding new actors to the group (Gomes-Casseres, 2001). Here one has to think of rising coordination costs and scarcity of management capacity, which can constrain further alliance formation (Gomes-Casseres, 2001). However these social factors related to the operational and managerial issues concerning the internal organization of alliance groups fall beyond the scope of this research.

Apart from the social component in network evolution and group formation specifically, there is also a technological driving force (Kash and Rycoft, 2000). Firms search for those technologies that enable them to extend their established technological capabilities (Stuart and Podolny, 1996). Hence they search for partners with whom they share technological content and with whom they are either directly or indirectly linked in the technological network (Podolny and Stuart, 1995). They search for those new technologies that can add to their existing technological base and which can be built upon in later periods. The ability to build on this technology is strongly related to their previous R&D activity (Rosenkopf and Nerkar, 2001), which requires pre-alliance technological overlap or absorptive capacity (for example Cohen and Levinthal, 1990; Hamel, 1991; Lane and Lubatkin, 1988; Mowery et al., 1996) to absorb the partners' technological capabilities (Tsai, 2001). This similarity in the partners' technology portfolio is required for the replication of the firms' ties. Hence alliance blocks within inter-organizational technology networks typically form among companies who are technologically similar.

1.6 ALLIANCE BLOCK FORMATION FROM A NETWORK EVOLUTIONARY PERSPECTIVE: THE SOCIO-TECHNICAL SYSTEM

From the above we can conclude that in the network evolutionary process, and in the subgroup formation process specifically, both social driving forces and technological motivations are important. However these driving forces do not develop in isolation from each other, but happen at the same time in a co-evolutionary way: social networks and technology development constantly shape each other. Thus in technology development or innovation of complex technologies in networks of relationships, the networks co-evolve with their technologies in so-called socio-technical systems. As technology development in networks is based on co-evolution in socio-technical systems, it involves both a network of interacting firms as well as a technology that moves along a trajectory or path embedded within that technological community (Kash and Rycoft, 2000). Hence trust within those communities allows members to build up social capital through interaction. This social capital gives them access to 'a stock of collective learning that only can be created when a group of organizations develops the ability to work together for a mutual gain' (Kash and Rycoft, 2000: 821). The technological community shares a particular body of knowledge and has 'broad agreement on the key technological and organizational obstacles and opportunities likely to be encountered in the future evolution of the trajectory' (Kash and Rycoft, 2000: 821).

1.7 RESEARCH QUESTIONS

Since the focus of our study is to further develop the theory on social structure and its effect on alliance formation patterns in social networks, we are interested in empirically identifying the social mechanisms that induce those network dynamics. These social mechanisms can cause enabling and constraining effects of embeddedness in alliance networks. Apart from the network enabling effect of embeddedness (for example Granovetter, 1992; Gulati, 1998) in alliance network formation and in alliance block formation in particular – where embeddedness is a driving factor in the network evolutionary process – the academic literature has given considerably less attention to the constraining effect of embeddedness in the decision about who to partner with. In many cases the enabling effect of embeddedness in alliance formation that is based on preferential relations can turn into a paralysing effect as actors become locked-in. In that situation actors only rely on partners in their own closed social system. Those firms may start to suffer from 'over-embeddedness' (Uzzi, 1997) in technological terms as well as in relational terms. Then, through the replication of ties in the group, firms in cohesive subgroups tend to become more similar and 'relationally inert' (Uzzi, 1997; Gargiulo and Benassi, 2000). The latter is also known as strategic gridlock (Gomes-Casseres, 1996) which forces firms to exclude attractive partners, as they have become unavailable, because they partnered with other groups. This is likely to put a severe strain on their ability to move flexibly into other 'resource niches'. The constraining effect of embeddedness can lead to decreasing opportunities for learning and innovation for block members involved. In turn it can cause block dissolution as members look for partners outside of the block to get access to new information to speed up their opportunities for innovation.

From this we can conclude that there are specific social mechanisms that cause this dual effect of embeddedness in the dynamics of inter-organizational networks. Therefore we intend to empirically illustrate this dual effect of embeddedness by addressing those social mechanisms that cause these enabling and constraining effects. This leads to the following research question:

1. What is the role of embeddedness and its social mechanisms in the dynamics of inter-organizational networks and alliance blocks specifically?

Next to the role of embeddedness (for example Granovetter, 1992; Gulati, 1998) as a driving factor in the network evolutionary process of alliance subgroup formation, we are interested in the effects of embeddedness on the innovative performance of these firms involved. This is a debated issue, because despite the growing stream of theoretical contributions, the academic literature and empirical research is rather indecisive about the performance effects of group membership. In order to begin filling this void we empirically examine the relation between network positioning – in terms of alliance group membership – and innovative performance. Thus:

2. What is the effect of embeddedness – alliance block membership in particular – on innovative performance?

Here, we have to acknowledge the relevance of group-based technology competition (Gomes-Casseres, 1996) in high-tech sectors. This form of competition affects the technology-positioning strategies at the firm level and hence the innovative performance of the firms involved, as they choose to either be an alliance block member or not. Alliance block membership can be understood as one of the strongest forms of social embeddedness. The effect of block membership on the innovative performance of those companies can therefore be seen in the light of the current debate on the advantages and disadvantages of social embeddedness (for example Burt, 1992; Coleman, 1988; Rowley et al., 2000; Gargiulo and Benassi, 2000). A network position as a block member in a dense and closed part of the alliance network in a closely connected subgroup provides a strong basis of trust and intimacy for the companies involved. Trust ameliorates information sharing, reduces resistance and provides comfort amongst the partners (Krackhardt, 1992; Brass et al., 1998; Granovetter, 1973). These densely connected blocks of technology alliances are probably well able to facilitate the essential knowledge transfer to innovate. In that case embeddedness enables the transfer of knowledge among the partners. However as a result of the growing similarity of companies collaborating solely in these alliance groups, the development of radical innovations in these groups can be hampered.

Moreover we would like to improve our understanding of how firms should position themselves in strategic alliance networks under various technological conditions, in order to maximize their innovative performance (Figure 1.2). With technological conditions we refer to the degree of turbulence in the technological environment that is either structure-reinforcing or structure-loosening (Madhavan et al., 1998), caused by respectively incremental technological developments or disruptive technologies (for example Bower and Christensen, 1995). Therefore we pose the following question:

3. How does technological change – that is either disruptive or incremental – mediate the relationship between block membership and innovative performance?

To answer these research questions we will develop and empirically test hypotheses, in order to increase our understanding of network dynamics and innovation.

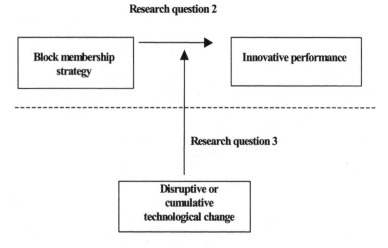

Figure 1.2 Conceptualization of research questions 2 and 3

1.8 OUTLINE OF THE STUDY

In the second chapter of this book, we will discuss the definitions of strategic technology alliances and alliance blocks and we will also provide a description of the data and methodology used in this book.

Furthermore we will describe the variables concerning the empirical part of this study. In addition we will address the major technological developments and trends in the industry of our focus: the microelectronics industry from 1970 to 2000.

The first research question will be addressed in the third and fourth chapters in which we will describe the role of embeddedness in the evolution of alliance networks and cohesive subgroups specifically. By introducing our first propositions, these chapters will empirically illustrate the social mechanisms that enable and constrain alliance block formation.

In Chapter 5 we will address the specific attributes of alliance block members compared to non-block members. This insight is useful as these characteristics have implications for the relation between block membership and innovative performance.

In Chapter 6 we will empirically study our second research question on alliance block membership and innovative performance, where we will derive some basic hypotheses on the effect of alliance block membership on innovative performance.

In Chapter 7 we will empirically test some basic hypotheses on the effect of block membership on innovative performance under changing technological conditions, which will answer our third research question.

In Chapter 8 we will discuss, conclude and reflect on the outcomes of our hypothesis testing and on the research questions. Finally we will state the limitations of the study and give directions for further research.

Chapter 9 will give a short summary of the book.

NOTES

1. In this book we use alliance blocks, (technology-driven) constellations, cliques and (alliance) (cohesive) (sub)groups interchangeably.
2. The relevance of this topic, as for instance demonstrated by the growing importance of strategic technology alliances as a major element in the external linkages of companies, has been documented in many publications (Hagedoorn and Duysters, 2002). See Hagedoorn (1996) and Osborn and Hagedoorn (1997) for an overview of the literature.

2. Definitions, Data and Methodology

2.1 INTRODUCTION

In this chapter we will discuss the definitions of strategic technology alliances and alliance blocks. Furthermore we will give a description of the data and methodology we used for the empirical part of the study and describe the variables we used in the analyses. And we will address the major technological developments and trends in the industry of our focus: the microelectronics industry from 1970 to 2000.

2.2 DEFINITIONS

Strategic technology alliances are defined as the establishment of common interests between independent (industrial) partners, which are not connected through (majority) ownership (Hagedoorn, 1993). And: 'cooperative agreements aimed at joint innovative efforts or technology transfer that can have a lasting effect on the product-market positioning of participating companies' (Hagedoorn and Schakenraad, 1994: 291). Others define strategic technology alliances as 'any organizational structure used to govern an incomplete contract between separate firms and in which each partner has limited control' (Gomes-Casseres et al., 2002: 4). Or 'cooperative efforts in which two or more separate organizations, while maintaining their own corporate identities, join forces to share reciprocal inputs' (Vanhaverbeke et al., 2002: 715).

Following from the definitions formulated above, we adhere to the notion that strategic technology alliances are cooperative agreements for reciprocal technology sharing and joint undertaking of research, between independent actors that keep their own corporate identity during the collaboration. Although alliances can cover a number of activities, such as marketing, production, distribution and R&D, we decided to focus on technology or R&D alliances. Typical examples of strategic technology alliances are joint research pacts, joint development agreements, R&D contracts, (mutual) second sourcing agreements, cross-licensing, research corporations, R&D joint ventures and technology-inclined cross-holdings. The differentiation between strategic alliances and other forms of cooperation is described extensively in Hagedoorn (1993).

Networks refer to inter-organizational relationships involving strategic technology alliances and can range from sparse dyadic, to dense multilateral relations where actors tend to cluster in alliance blocks.

Technology alliance blocks or technology-driven groups are defined as multiple partner firms linked with each other through strategic alliances into groups or constellations (Das and Teng, 2002) 'competing against other such groups and against traditional single firms' (Gomes-Casseres, 1996: 3). These firms are bound together by a network of relatively strong ties (Vanhaverbeke and Noorderhaven, 2001). Hence we define technology alliance blocks as groups of firms connected through strategic technology alliances with the purpose of joint innovative effort, where alliance block members maintain and replicate strong and multiple ties within their group as compared to the outside. Thus, the main characteristic of an alliance block or a cohesive subgroup in a social network is that the relationships among its members are more important and more numerous than the relations between members and non-members (Fershtman, 1997). There are four ways to conceptualize the idea of subgroups. These properties address the mutuality of ties (adjacency), the closeness or reachability of subgroup members, the frequency of ties among members, and the relative frequency of ties among block members compared to non-members. Following this classification we can discern several cohesive subgroup measures like cliques concerning the mutuality of ties; n-cliques, n-clans and n-clubs to address the reachability of group members; k-cores and k-plexes to specify the frequency of ties among members; and LS sets and lambda sets to focus on the relative frequency of ties among members compared to non-members (Wasserman and Faust, 1994).

A clique is maximal complete subgraph of three or more nodes. These nodes are all adjacent to each other. There are no other nodes that are also adjacent to all of the members of the clique (Wasserman and Faust, 1994). However, since some of the above mentioned measures like cliques and n-cliques are quite stringent for data analysis, and since we are interested in conceptualizing block membership as members having more numerous relations with each other than with non-block actors (Alba, 1973), we measure a cohesive subgroup in a social network by comparing 'the prevalence of ties within the subgroup to the sparsity of ties outside the subgroup' (Wasserman and Faust, 1994: 267). This means we are interested in the relative frequency of ties among block members compared to non-members (see classification above). Therefore we have to compare the properties of ties among members within the subgroup to properties of ties to actors outside of the subgroup (Wasserman and Faust, 1994).

2.3 DATA SOURCES

All the data regarding the strategic technology alliances were taken from the MERIT-CATI database (Duysters and Hagedoorn, 1993). The CATI databank contains information on thousands of cooperative technology agreements and their 'parent' companies and currently covers the period between 1970 and 2000 containing information on nearly 15000 alliances of parent companies active in biotechnology, information technology, new materials and a number of 'non-core' technologies. Within the CATI database there are 65 classifications with respect to sub-sectors and sub-fields of technology that can also be identified in terms of SIC codes. These industries do not only refer to high-tech industries but also to sectors such automotive, chemicals, metals, machinery, food and beverages. The most important data sources are international and specialized trade and technology journals for many fields of technology and industrial sectors. The alliances in the database are primarily related to technology cooperation. In this specific form of alliances the transfer of technology or the joint undertaking of research has to be part of the agreement: for example joint research pacts, joint development agreements, R&D contracts, (mutual) second sourcing agreements, cross-licensing, research corporations, R&D joint ventures and technology-inclined cross-holdings. Mere production or marketing alliances are excluded. The strategic nature of these alliances is safeguarded by strict inclusion rules, which ensure that only those alliances are included that are undertaken in order to affect the long-term strategic positioning of companies. See Hagedoorn and Schakenraad (1994) for a further description of this data bank.

Although these technology partnerships constitute only a small group out of the overall population of inter-firm partnerships, they are accepted in the literature as more than useful indicators of the behaviour of companies with regard to their inclination to cooperate with others (for example Hagedoorn, 1996; Kogut, 1989; Mowery, 1988; Mytelka, 1991). For the purpose of the present analysis, information was used regarding the industrial sectors in which companies operate, the year of establishment of the technology partnership and its industry affiliation, the number and names of partners involved and the modes of cooperation.

Regarding our innovative performance variable, we obtained data from the US Patent and Trademark Office database (US Department of Commerce) for the period 1980–2000. Although we are aware of the potential bias in favour of US companies against non-US companies in this database, the innovation literature suggests that US patents are a valid indicator to use in innovation research. In particular the attractiveness, importance and level of sophistication of the US market, in combination with the amount of patent protection, makes it almost a necessity for non-US companies to file patents in the US (Patel and Pavitt, 1991; Hagedoorn and Duysters, 2002).

The innovative performance of companies is not expected to be only dependent on the specific network position of companies, but also on firm-specific characteristics. In that context one has to think of the size of companies that captures scale and scope effects, home regions of firms and technological specialization and R&D efforts, which might generate differentials in technological performance. We control for these aspects in our analyses. Therefore information required for our control variables, like the revenues and R&D expenditures, was accessed through well-known databases such as Compustat, Disclosure, Securities Data and Worldscope. We took specific company characteristics, such as the size of companies, country of origin information and so on, from Van Dijk's global researcher files.

2.4 METHODOLOGY

This quantitative research project will be primarily based on an extensive literature study and the empirical analysis of the CATI alliance database (Duysters and Hagedoorn, 1993). As this research project is theory-oriented and deductive in nature, we will derive hypotheses that will be tested empirically. The hypotheses of this study are statements about the effect of embeddedness and its social mechanisms on alliance network formation, and block formation in particular. Furthermore our hypotheses address the relation between patented innovation output and the firm's network positioning under changing technological conditions. The extensive CATI alliance database enables us to explore the historical patterns of the formation of alliance networks, and in particular in the formation of alliance blocks. In order to test our hypotheses we computed several social network measures by constructing adjacency matrices representing the relationships among the firms in the strategic technology alliance network. Various network measures like alliance block membership were calculated using UCINET (Borgatti et al., 1999, 2002). Furthermore we used software from the Centre for Global Corporate Positioning (CGCP) to plot the network graphs.

The first major research implication stems from the further theoretical development of the concept of alliance blocks or 'technology driven constellations' (Gomes-Casseres, 1996) by taking a social network perspective (for example Nohria, 1992; Gulati, 1998). From this, progress can be made toward empirical research on these alliance blocks based on social network measures (for example Wasserman and Faust, 1994). Therefore the first step in the empirical part of this project is concerned with the detection of alliance blocks in the microelectronics alliance network. Formal network analysis tools can be used to detect so-called cohesive subgroups (alliance blocks) in the networks.

The second research implication involves demonstrating the duality of embeddedness in the network evolutionary process by identifying the social mechanisms that cause this duality. In the beginning of the alliance network formation process, embeddedness can be seen as an enabling factor in alliance formation. However as the alliance network formation process proceeds, it can create a paralysing effect in the alliance blocks. It can even become a liability for the alliance block members involved. By measuring the size of the network in consecutive periods, and the number of groups and group members in those periods, we should be able to pass judgement on the enabling effect on embeddedness. Furthermore by composing technology profiles of the block members involved and by calculating the in-group/out-group ratios as a measure for replicating ties in the group, we are able to address the constraining effects of embeddedness.

The third major research implication points out the relation between alliance block membership and innovative performance, where alliance blocks can be seen as competitive weapons to establish and sustain technological advantages. For an assessment of the determinants of success of the firms pursuing an alliance block strategy, we use a combination of social network analysis measures and 'standard' technology indicators like R&D intensity and technological specialization. The innovative performance of group members can be measured by means of the patent intensity of group members. We use standard multivariate statistical techniques to assess the determinants of the innovative performance of firms in this high-technology industry.

The fourth research implication addresses the moderating impact of the changing technological environment on innovative performance for alliance block members. Therefore we introduce an interaction effect based on the years one is pursuing an alliance block membership strategy, and the nature of technological change. The latter is based on the relative differences in technology profiles at the industry level.

We used a three-year moving window (Table 2.1) to include the previous history of the collaborations (see for example Koka and Prescott, 2002) and to indicate the average duration of a strategic technology alliance. This implies that we introduce a time lag of, on average, three years for joint innovative input, such as joint R&D projects, to materialize into innovative output, that is patents. Research on such time lags suggests that on average an invention leads to patents after about two and a half years, although there is substantial variation (Hagedoorn and Duysters, 2002). If we include the process of R&D itself and the additional time that joint projects can take, then an average time lag of three years appears to be a valid estimate (Hagedoorn and Duysters, 2002). We expect that block membership in period $t = 1$ will have an effect on innovative performance at $t = 4$, so we used a time lag of three years to measure innovative performance as a result of block membership (Table 2.1). For the dependent variable (innovative performance) we take the patents applied for during the period 1983–2000. For the independent variable block membership, we take the period 1980–97, introducing this average time lag of three years. This means that because of our three-year time lag, block membership in the period 1995–97 ($t = 16$) is the last period we use, corresponding to the number of patent applications in the period 1998–2000 ($t = 19$). This means we use 16 sets of data in our dataset covering the period 1980–2000 in microelectronics. Thus set one contains the alliances formed in 1980–82. Our second set involves the alliances formed in 1981–83, our third set contains alliances formed in 1982–84 and so on, and our sixteenth set contains alliances formed in the period 1995–97.

Table 2.1 Overview of variables used in the analysis

DEPENDENT VARIABLE	INDEPENDENT VARIABLE		MODERATOR VARIABLE (K1)		CONTROL VARIABLES		
Innovative performance (Y) (patent intensity)	Alliance block membership (X1) (lambda sets level 4)	Years in block (X1a)	Technological change (Z1)	Years in block (X1a)	Country (dummy) (X2), (X3)	R&D intensity (X4) (R&D expenses / revenues)	Technological specialization (X5) (patents in microelectronics / total patents)
1983–85 (t4)	1980–82 (t1)	1980–82 (t1)	0.13471	1980–82 (t1)			1980–82 (t1)
1984–86 (t5)	1981–83 (t2)	1981–83 (t2)	0.10272	1981–83 (t2)			1981–83 (t2)
1985–87 (t6)	1982–84 (t3)	1982–84 (t3)	0.10272	1982–84 (t3)			1982–84 (t3)
1986–88 (t7)	1983–85 (t4)	1983–85 (t4)	0.10272	1983–85 (t4)			1983–85 (t4)
1987–89 (t8)	1984–86 (t5)	1984–86 (t5)	0.23492	1984–86 (t5)			1984–86 (t5)
1988–90 (t9)	1985–87 (t6)	1985–87 (t6)	0.23492	1985–87 (t6)			1985–87 (t6)
1989–91 (t10)	1986–88 (t7)	1986–88 (t7)	0.23492	1986–88 (t7)	USA		1986–88 (t7)
1990–92 (t11)	1987–89 (t8)	1987–89 (t8)	0.00000	1987–89 (t8)	EUROPE		1987–89 (t8)
1991–93 (t12)	1988–90 (t9)	1988–90 (t9)	0.00000	1988–90 (t9)	ASIA		1988–90 (t9)
1992–94 (t13)	1989–91 (t10)	1989–91 (t10)	0.00000	1989–91 (t10)		1989–91 (t10)	1989–91 (t10)
1993–95 (t14)	1990–92 (t11)	1990–92 (t11)	0.26979	1990–92 (t11)		1990–92 (t11)	1990–92 (t11)
1994–96 (t15)	1991–93 (t12)	1991–93 (t12)	0.26979	1991–93 (t12)		1991–93 (t12)	1991–93 (t12)
1995–97 (t16)	1992–94 (t13)	1992–94 (t13)	0.26979	1992–94 (t13)		1992–94 (t13)	1992–94 (t13)
1996–98 (t17)	1993–95 (t14)	1993–95 (t14)	0.06619	1993–95 (t14)		1993–95 (t14)	1993–95 (t14)
1997–99 (t18)	1994–96 (t15)	1994–96 (t15)	0.06619	1994–96 (t15)		1994–96 (t15)	1994–96 (t15)
1998–00 (t19)	1995–97 (t16)	1995–97 (t16)	0.06619	1995–97 (t16)		1995–97 (t16)	1995–97 (t16)

2.5 VARIABLES

Because of our preoccupation with studying the innovative performance of alliance block members, we decided to focus on companies as the level of analysis in this study. The main reason for taking companies as our main object of study is that innovative performance can be traced back at the company level and not at the level of an individual alliance. The effects of small individual alliances might be difficult to trace back, whereas the combined effect of a number of alliances is easier to detect. Furthermore patents are applied for by companies and not at the level of the alliance (Hagedoorn and Duysters, 2002). We use a single-industry design to deal with industry effects, which can affect a firm's propensity to patent.

Dependent variable
We measure the dependent variable innovative performance by taking the patent intensity of firms (Duysters and Hagedoorn, 2001; Hagedoorn and Duysters, 2002), that is the number of patents applied for divided by firm size (revenues) in the period 1983–2000.[1] There are some indications in the literature that larger companies have a higher propensity to engage in partnerships than smaller companies (Duysters and Hagedoorn, 1995; Mytelka, 1991). We have chosen revenues as an indicator for firm size instead of the more frequently applied employment indicator to account for the effects of quasi-integration. Japanese companies often have fewer employees than their US and European competitors on account of the Japanese lean production methods and sophisticated customer supplier networks (Duysters and Hagedoorn, 1995). Size in terms of revenues is therefore in our opinion an appropriate indicator of economic magnitude to compare companies from different regions (Duysters and Hagedoorn, 1995).

The use of patent statistics has been criticized on many occasions (Levin et al., 1987; Cohen and Levin, 1989; Griliches, 1990; Archibugi, 1992). Criticism focused on the use of patent counts as a dependent firm-level performance indicator because the link between changes in patent activity and a specific sourcing relationship is tenuous as firms pursue multiple alliances simultaneously (Steensma and Corley, 2000). Moreover patents represent codified knowledge and hence do not measure intangible benefits gained from technology sourcing (Steensma and Corley, 2000). And only some inventions are suited to patenting, others are not patented and the patented ones can differ greatly in economic value (Cohen and Levin, 1989; Griliches, 1990; Trajtenberg, 1990; Ahuja and Katila, 2001). Related to that, these problems depend on the kind of research and appropriability policies (Cohen and Levin, 1989; Levin et al., 1987; Ahuja and Katila, 2001). Furthermore the tendency to patent has been found to differ across industries (Stuart and Podolny, 2000).

However despite these shortcomings there are some strengths related to patents as a measure of innovation output. Patents are directly related to inventiveness of a company (Ahuja and Katila, 2001) and they represent an externally validated measure of technological novelty (Griliches, 1990; Ahuja and Katila, 2001). Patents have economic significance as they confer property rights on the assignee (for example Ahuja and Katila, 2001). Patents also connect well with other measures of innovative output (Ahuja and Katila, 2001), like new products (Comanor and Scherer, 1969), innovation and invention counts (Achilladelis et al., 1987; Ahuja and Katila, 2001) and sales growth (Scherer, 1965). Furthermore this indicator is generally considered to be the most appropriate measure of innovative performance at the company level (Acs and Audretsch, 1989; Hagedoorn and Duysters, 2002) that enables us to compare the innovative performance and technological learning of companies (Acs and Audretsch, 1989; Aspden, 1983; Hagedoorn and Duysters, 2002) especially in a single-industry high-tech sector study. Limiting the study to a single industrial sector minimizes problems related to other factors affecting patent propensity as these factors are likely to be stable within one context (Basberg, 1987; Cohen and Levin, 1989; Griliches, 1990; Ahuja and Katila, 2001).

In all, this patent indicator is particularly relevant for our study of networks of strategic technology alliances (Hagedoorn and Duysters, 2002) and hence for investigating groups of strategic technology alliances, which influence the technological learning capabilities of individual companies. Patents are thus 'signals' of technological competencies and learning capabilities of companies in inter-firm networks (Powell and Brantley, 1992; Hagedoorn and Duysters, 2002.)

Independent variable
Our independent variable alliance block membership or cohesive subgroup membership refers to pursuing a network position in an alliance block. In operationalizing this construct we do not make a distinction between the several roles alliance block members can

occupy, like for example core or periphery players in the alliance block (Everett and Borgatti, 1999a, 1999b). The same holds for non-alliance block members, as we do not distinguish between brokers occupying a structural hole position or non-alliance block members occupying a peripheral position in the network. In other words we put all non-alliance block members' roles in one group (non-alliance block membership) and we lump all alliance block members' roles together (alliance block membership).

Alliance blocks we define as cohesive subsets of (similar) actors in a network based on their relative inward to outward interactions (Fershtman, 1997; Knoke and Kuklinsky, 1982). Hence in this thesis we use the hierarchical clustering measure lambda set that fits this idea of comparing in-group ties to out-group ties. Lambda sets consider the line connectivity of subgroup members compared to non-group members (Wasserman and Faust, 1994). To be more specific, a lambda set should be hard to disconnect by the removal of lines from the subgraph.[2] Line connectivity indicates the extent to which a pair of nodes remains connected by some path (denoted as $\lambda (i,j)$), even when lines are deleted from the graph. Thus, the line connectivity of two nodes i and j is equal to the number of paths between them that contain no lines in common. Based on the property of line connectivity Borgatti et al. (1990) define a lambda set as follows: 'The set of nodes Ns, is a lambda set if any pair of nodes in the lambda set has larger line connectivity than any pair of nodes consisting of one node from within the lambda set and a second node from outside the lambda set.' The smaller the value of $\lambda (i,j)$, the more vulnerable i and j are to being disconnected by removal of lines. The larger the value of $\lambda (i,j)$, the more lines must be removed from the graph in order to leave no path between i and j (Wasserman and Faust, 1994). The algorithm employed first computes the maximum flow (that is the connectivity) between all pairs of vertices and uses this information to construct the lambda sets (Borgatti et al., 2002).

These lambda sets do not overlap, unless one is contained within another. Another important property of lambda sets is that nodes within a lambda set are not necessarily cohesive in terms of either adjacency or geodesic distance. Thus, members of a lambda set do not need to be adjacent, and since there is no restriction on the length of paths that connect nodes within a lambda set, members of a lambda set may be quite distant from another in the graph (Borgatti et al., 1990). Furthermore, lambda sets do not measure cohesiveness based on in-group versus out-group strength; they measure relative frequency of ties among members compared to non-members (Wasserman and Faust, 1994).

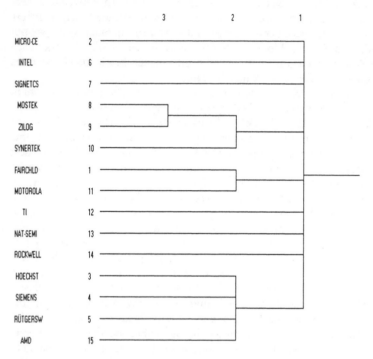

Figure 2.1 Line connectivity in 1975–77

To measure subgroup cohesion specifically as in-group versus out-group strength, we introduce a measure that incorporates this cohesiveness in an in-group/out-group ratio similar to the one mentioned in Wasserman and Faust (1994). Here the ratio of the strength of ties within the subgroup to ties between subgroups does not decrease appreciably with the addition of new members (Wasserman and Faust, 1994).

In our analysis we measure block membership by calculating lambda sets at the hierarchical clustering levels two and four. Those levels indicate that a group at level two is less densely tied than a group at level four. We have chosen those two critical values, because those are the levels we have available for all our datasets from 1980 to 2000. Figures 2.1, 2.2 and 2.3 show the hierarchical clustering dendrograms. The level at which any pair of actors is aggregated is the point at which both can be reached by tracing from the start at the right of the figure to the actors at the left of the figure (for example Figure 2.2). The scale at the top indicates the level at which they are clustered (Borgatti et al., 2002). Each level corresponds to a different degree of minimum internal line-connectivity. This value characterizes the lambda set. In period 1976–78 we can discern the first lambda sets at level four (compare Figures 2.1 and 2.2).

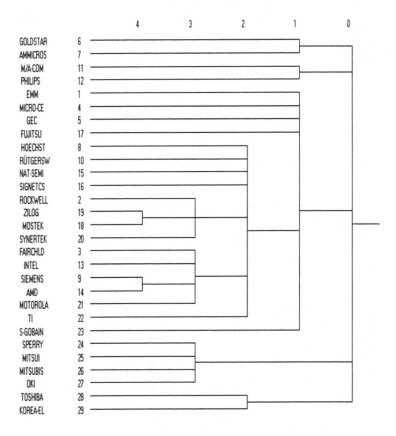

Figure 2.2 Line connectivity in 1976–78

The first group consists of the companies Mostek and Zilog(Zilog Inc.) and the other set is Siemens (Siemens A.G.) and AMD (Advanced Micro Devices Inc.). In our analyses, we assign a dummy

1 for block members present in a lambda set and a dummy 0 for non-block membership at both the hierarchical clustering levels. As more actors come into the network, the levels can go up to 24 (Figure 2.3).

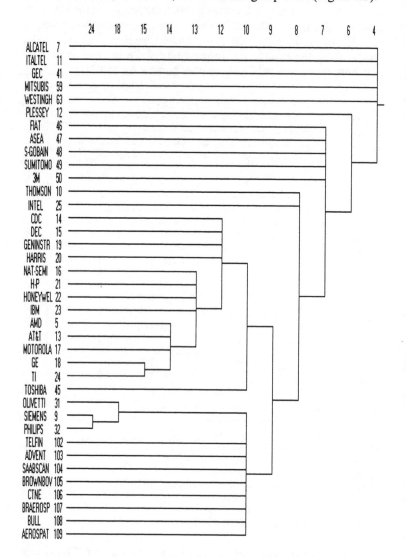

Figure 2.3 Line connectivity in 1985–87

Control variables
We use three control variables in our analyses. We control for the national background of companies, since there might be different propensities to patent and to undertake alliances for the various regions. We distinguish three home regions – the United States, Europe and Asia.

Furthermore as we study technology partnerships, we will control for the R&D intensity; the ratio of microelectronics-related R&D expenditures to revenues. We expect a positive effect of R&D intensity on patent activity, as these research efforts will (at least partly) be transformed into patents (Hagedoorn and Duysters, 2002). The latter might be closely related to the innovative performance of companies. Thus studies have found a positive relationship between R&D input and technological output, measured in terms of patents (Hagedoorn and Duysters, 2002). This relationship might however not be linear as patenting output may decrease gradually with an increase in R&D expenditure (Hagedoorn and Duysters, 2002).

Additionally we incorporate the control variable of technological specialization, where we divide the patents applied for in the possible 15 semiconductor classes by the total amount of patents applied for per company per period. We expect that companies, which have a higher number of patent applications in microelectronics (higher specialization in microelectronics) as compared to their overall patent applications, will be more innovative than firms that are less specialized in microelectronics.

2.6 THE MICROELECTRONICS INDUSTRY

History
The roots of microelectronics can be traced back to the turn of the century and Lee DeForest's grid vacuum tubes. Today's technologies come most directly from the transistor, developed at the Bell laboratories in 1947 (Duysters, 1996). Silicon is the most widely used semiconductor material used these days, where gallium arsenide is also becoming very important. Semiconductor devices are at the heart of electronics. The development of integrated circuits and microprocessors has induced denser and more complex versions of those initial transistors (Duysters, 1996). This transition from transistors to the integrated circuits led to the disappearance of the vertically integrated American electronics companies who were leaders in the production of vacuum tubes and had managed to stay in

the race during the discrete semiconductor (transistors) era (Langlois and Steinmüller, 2000). Market shares declined as new entrants invaded the market and specialized manufacturers like TI, Fairchild and Motorola started to grow (Langlois and Steinmüller, 2000).

Characteristics of the industry
The microelectronics industry is highly volatile, subject to constant challenge and change, as markets change as swiftly as the technologies involved. Innovation is a continuous survival requirement in this industry. The microelectronics industry comprises processors, accelerator chips, RISC-processors, memory chips (ROM, EPROM, dRAM – dynamic random access memories –, sRAM), peripheral chips, ASICs, expansion and other chipboards and transistors. These technological groups imply a substantial diversity, but within the microelectronics industry, semiconductors are the main underlying technology.

The industry is becoming increasingly complex. One issue in this complexity is the extremely broad spectrum of the microelectronics knowledge base as a result of the vast array of its applications and derivative technologies. Semiconductor technology is cumulative, and domain-specific as a result of its complexity (Podolny and Stuart, 1995). Production technology for semiconductor products is easily the most complex process ever adapted to mass production. Another factor that contributes to the complexity in this industry is the amount of fierce competition. Competition unfolds simultaneously on technological advance and on price. This complexity requires high R&D expenses as the production technology becomes more complex too. This results in a capital-intensive industry staffed by highly trained specialists and characterized by a scientific approach to manufacturing (Jelinek and Schoonhoven, 1990).

Semiconductor production has grown more demanding and more expensive as the requirements for control, precision and complexity have grown. This implies that firms have to incur huge capital investments, of which the costs must be amortized across large volumes of production, and thus through economies of scale. However because of the fast-changing pace of technologies and dynamic competitive environment, time to recoup the heavy investments is actually too short. Furthermore, the equipment must be able to produce many components, products and designs in economies of scope (Jelinek and Schoonhoven, 1990). This means that in order to attain volume production, economies across a broad scope of designs have to be achieved. Thus volume is now accomplished across a related product family instead of on a single product as in the past (Jelinek and Schoonhoven, 1990). In this way the cumulative volume of production increases and costs decrease, which permits price reductions.

The firms active in the semiconductor industry can either be merchant semiconductor producers or captive producers. With the former we refer to semiconductor producers that sell their products on the open market, like for example Intel and TI. With captive producers we refer to firms that produce more than 75 per cent of their products for their own high-tech products, for example IBM, or for universities or national governments or agencies (Podolny et al., 1996).

Since semiconductor production involves complex processes (Stuart and Podolny, 1996) as explained above, technical advances have driven down the price and increased the performance of semiconductor devices throughout the history of the industry. For this reason R&D expenditures are high (exceeding 10 per cent of revenue for many incumbents), and firms' decisions about which technological areas to target are critical factors in determining organizational performance (Podolny and Stuart, 1995).

Thus intense and global competition, complex and science-based production processes, technological obsolescence and technological innovation at high rates characterize this industry (Jelinek and Schoonhoven, 1990). Concerning production, economies of scope have to be reached across broad product families through learning curve pricing.

The number of strategic alliances is increasing gradually in this industry (Figure 2.4).

Motivations for choosing the microelectronics industry

Our empirical analysis covers the industrial, technological and networking activity of companies operating in the international microelectronics industry. There are several reasons for choosing this particular industry and its network of strategic technology alliances.

The industry has been technology-driven throughout its history, which indicates that technological positioning and strategy are keys to survival (Podolny and Stuart, 1995; Stuart and Podolny, 2000).

The microelectronics sector is a typical high-tech sector that is driven by technology competition. It is an industry where one finds a large number of strategic technology alliances that play an important role in the competitive strategies of companies (see amongst others,

Duysters and Hagedoorn, 1998; Gomes-Casseres, 1996; Hagedoorn and Schakenraad, 1992; Mytelka, 1991).

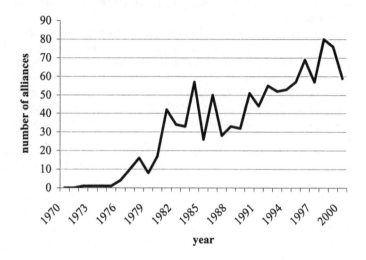

Figure 2.4 The number of alliances formed in microelectronics

It is well documented that alliances are an important element of the technology acquisition strategies of companies in high-tech sectors (for example Stuart and Podolny, 1996, 2000; Hobday, 1997; Langlois and Steinmüller, 2000; Vanhaverbeke et al., 2002; Rowley et al., 2000; Holbrook et al., 2000; Stoelhorst, 2002; Park et al., 2002; West, 2002). Many of the alliances that incumbents have formed (see Hagedoorn, 1993) are horizontal relationships between semiconductor firms aimed at new technology development (Stuart and Podolny, 2000). Although strategic technology alliances may also play a role in other sectors, the effect of network positioning or alliance block membership on technological performance is probably most evident in high-tech sectors.

Furthermore the microelectronics industry is a strategically important sector. It can be seen as the driving force of technological change in virtually all sectors of the information technology industry. It is of strategic importance not only in terms of market size, but also because its outputs are vital components in a wide range of other products. As microelectronics play an increasing role in related industries such as computers, systems and peripherals, the diversity and the central role of microelectronics are becoming clearly visible elsewhere (Jelinek and Schoonhoven, 1990).

Finally, the industry has a high propensity to patent, especially in the period of our study. This allows us to track the innovative performance of the companies in our sample by means of their patent activity. Moreover the microelectronics sector and its companies are very well documented in terms of available company information and sectoral data.

NOTES

1. We summed up the patent applications per firm per year for our sample of 135 companies in the 15 microelectronics patent classes in periods of three years. Thus period 1 is 1983–85, period 2 is 1984–86, period 3 is 1985–87 and so on. Then we divided the number of patents applied for by the revenues, in order to correct for firm size. We measure size of companies by taking the revenues that companies realized during the period 1989-97.
2. See Wasserman and Faust (1994), p. 270.

3. The Enabling Effect of Embeddedness[1]

3.1 INTRODUCTION

This chapter will be one of the first descriptive empirical attempts to study the dynamics of alliance networks from a longitudinal perspective. We will address our first research question, where we will describe the role of embeddedness and social capital in the process of alliance block formation in technology alliance networks. By introducing our first propositions, this chapter will empirically study the main social mechanisms like local search and replication of ties that create the enabling effect of embeddedness and drives alliance block formation. Then, through the formation of subsequent ties, firms in social systems tend to rely heavily on their direct and indirect contacts in forming new partnerships. This so-called 'local search' enables firms to create trustworthy and preferential relations. Over time those relations tend to develop into strong ties, as firms rely on the same partners when replicating their existing ties; this constitutes the enabling effect of embeddedness. We develop two propositions that address the enabling effects of embeddedness in the worldwide microelectronics industry from 1970 to 2000. We illustrate these propositions by means of descriptive empirical evidence.

3.2 THEORETICAL BACKGROUND

Social networks are the embedded social relations that surround the actors in the alliance network and indicate how these actors are connected and related. By investing in these social relations through the replication of their existing ties, firms build up social capital. Social capital relates to the investment in social relations that generates expected returns (Lin, 1999). It is defined as 'the sum of resources that accrue to a firm by virtue of possessing a durable network of relationships' (Bourdieu and Wacquant, 1992: 119; Koka and Prescott, 2002). Thus social capital refers to the potential beneficial network of relations with external parties as well as the resources embedded in that network that may be accessed and mobilized for purposive actions (Lin, 1999; Burt, 1992; Nahapiet and Ghosal, 1998; Chung et al., 2000). This capital creates an advantage for individuals or groups in attaining their goals, as their interconnectedness gives them access to certain resources embedded in the network which results in higher returns (Burt, 2000). Thus in the literature we find consensus that investing in social relations and hence improving connectedness enables accessing and using the resources embedded in those social networks, and results in gaining returns (for example Bourdieu, 1986; Coleman, 1988, 1990; Lin, 1999).

Most of the literature on social capital has taken a firm-level perspective. However in order to describe the full dynamics of group formation in social networks, the effect of social capital at the group level also has to be taken into account (Lin, 1999). Social capital at the group level refers to aggregation of individual returns that benefits the collective (Lin, 1999). Most of the literature on this subject focuses on how certain groups develop and maintain their social capital as a collective asset and how such a collective asset enhances group members' life chances (Bourdieu, 1986; Coleman, 1988, 1990; Putnam, 1993; Lin, 1999). Through dense or closed networks, collective social capital can be maintained and reproduction of the group can be achieved. Norms and trust play an important role in producing and maintaining the collective asset (Lin, 1999). Being part of a dense, cohesive and redundant network promotes a normative environment that involves trust and cooperation among its members (Coleman, 1988, 1990; Gargiulo and Benassi, 2000) and eventually leads to a situation of strong social cohesion within these subgroups in the network (Friedkin, 1984).

Below we will introduce propositions that address the social mechanisms that are based on the investment in social capital and social relations, which cause the enabling effects of embeddedness.

3.3 PROPOSITIONS

Enabling embeddedness: local search and replication of ties
Current alliance networks provide future alliance opportunities (Gulati, 1995a) and early participation may provide firms with

potentially valuable partnering possibilities for the future. Alliance-proactive firms in networks are therefore more likely to possess the specific knowledge related to the identification and the selection of appropriate alliance partners (Sarkar et al., 2001). Alliance pro-activeness is a first-mover advantage, as early mover firms tend to capture advantageous positions resulting from their partner choice. Thus pre-emption of valuable and scarce resources in partner space can be a source of strategic advantage (Dyer and Singh, 1998; Sarkar et al., 2001). As a result some partners are not available because they are already tied to the focal firm's competitors.

Since trust is an important basis for knowledge sharing and partner selection, firms tend to be locally biased in their search strategies (see for example Nelson and Winter, 1982; Cyert and March, 1963; Stuart and Podolny, 2000). They often engage in 'local search' in forming their subsequent ties. They tend to initiate new partnerships that share the context with the outcomes of prior searches.[2] In their technological positioning, firms thus search for those technologies that enable them to extend their established technological capabilities (Stuart and Podolny, 1996). They generally search for partners with whom they share technological content and with whom they are either directly or indirectly linked in the technological network. These preferential relations are path-dependent, as prior ties determine the formation of future linkages (for example Gulati, 1995a; Levinthal and Fichman, 1988; Walker et al., 1997; Tsai, 2000). Furthermore these ties ameliorate information sharing, reduce resistance and provide comfort among the partners.

Over time, partner attractiveness will remain high or becomes even stronger (Madhavan et al., 1998) and preferential relations tend to develop into strong ties, characterized by frequent interaction and heavy commitment to the relationship. Strong ties (Granovetter, 1973), are characterized by solid, reciprocal and trustworthy relationships. This type of relationship creates a large basis of trust and intimacy between the partners (Brass et al., 1998; Granovetter, 1973). As those firms replicate their preferential relations based on their social capital at the group level (Lin, 1999) and their embeddedness, the network self-generates and reproduces over time. Being embedded in a densely connected network as a result of a high amount of social capital, makes engagement in subsequent ties more likely (Walker et al., 1997). Social capital at the group level (Lin, 1999), in particular, is crucial in the process of alliance block formation as the network becomes denser. The network increasingly turns into a growing repository of information on the availability, competence and reliability of prospective partners (Walker et al., 1997; Gulati, 1995a; Powell et al., 1996). Thus when the size of the network grows – the number of actors increases – and actors form multiple relationships, the possibility increases that alliance blocks are formed. The latter results from actors who develop multiple cohesive ties through local search. We conclude that if the size of the network increases, alliance block formation is likely.

Proposition 1:
If the size of the network increases, alliance block formation becomes more likely.

Alliance blocks
Alliance block membership can be seen as one of the strongest forms of embeddedness. In the conceptualization of alliance blocks, there are four general properties that apply: the mutuality of ties, the closeness or reachability of subgroup members, the frequency of ties among members, and the relative frequency of ties among subgroup members compared to non-members (Wasserman and Faust, 1994). Specifically the number of ties an individual has within a group and the closeness of the entire group to outsiders matter (Wasserman and Faust, 1994). Alliance block members have more numerous or more intense relations with each other than with non-alliance block actors. Alliance blocks are generally characterized by highly cohesive subsets of similar actors in a network (Knoke and Kuklinski, 1982). Cohesion refers to the extent to which there is a relatively direct strong interaction among individuals in a social system, requiring only few intermediaries, that is, indirect links (Bovasso, 1996). Social forces operate through direct and indirect contacts among subgroup members and through the cohesion achieved within the subgroup, as compared to outside the subgroup (Wasserman and Faust, 1994). When actors have relatively frequent contacts (face-to-face) and when they are linked through intermediaries (Friedkin, 1984), greater homogeneity is expected.

The engagement in subsequent ties in these dense and cohesive parts of the networks can also be explained from a transaction-costs perspective. When information is lacking about the competencies and reliabilities of potential partners, developing a relation with a new actor involves uncertainty (Tsai, 2000). Hence firms invest a substantial amount of time and energy to establish strong relationships (Burt, 1992) through preferential partnering. However the commitment and specificity of investments required in the relationship generate sunk costs (Gomes-Casseres, 1996). Therefore changing transaction partners in the short run is not likely, since it involves significant switching costs and implies a risk that existing relationships will dissolve (Chung et al., 2000). Furthermore as actors develop 'specific routines for managing an interface with each other' (Gulati, 1995a: 626), they tend to become blind to new partnership opportunities and instead rely on previous partners and routines only (Tsai, 2000). Thus when trustworthy partners are readily available, searching for or switching to new partners is hard to rationalize in the alliance formation process (Chung et al., 2000). Therefore actors rather replicate their existing ties (Gulati, 1995a, 1998; Walker et al., 1997) through local search. They look for partners they are familiar with, and with whom they share similarities in technological content in their densely connected social system. In the past, several scholars have addressed the fact that social relations develop in a path-dependent way, in the sense that previous ties determine how the future relationships evolve (see for example Gulati, 1995a, 1995b; Levinthal and Finchman, 1988; Walker et al., 1997; Tsai, 2000).

Another reason why firms tend to replicate their existing partnerships is the danger of reputation effects. This fear deters firms in a web of relations from behaving opportunistically against each other, and it increases the stability and longevity of their alliance formation in their closed system. The likelihood that a firm acts unethically decreases when the firm is embedded in a network of relations, since this behaviour is communicated quickly to other partners in the network. Actors then update their evaluation of the opportunistic actor and may not trust or interact with that firm in the future, since the opportunistic actor violates the trust created at the network level as well as on the dyadic level (Rowley et al., 2000). Since unethical behaviour damages the reputation of the opportunistic firm, this becomes a critical issue in partner selection. These reputation effects prevent alliance block members from behaving unethically. Cutting ties in cohesive groups can also damage the reputation and can hamper the revitalizing of these severed ties in the future. Moreover reputation effects also hamper new tie formation in the group (for example Raub and Weesie, 1990; Gargiulo and Benassi, 2000).

For alliance block members, it may be difficult to maintain ethical norms regarding actors outside of their alliance blocks. The expression 'honour among thieves' (Brass et al., 1998) may be the result of strong and dense connections among the thieves, who do not hesitate to act unethically *vis-à-vis* outsiders, that is non-alliance block members. Furthermore alliance blocks may be more powerful in number and network position and therefore can afford to act unethically without fearing the consequences (Brass et al., 1998). This suggests the following hypothesis:

Proposition 2:
When looking for new partners, firms replicate their existing ties within the subgroup.

3.4 SAMPLE AND DATA

The data on strategic alliances and characteristics of companies involved in these alliances is derived from the MERIT-CATI databank on strategic technology alliances (Duysters and Hagedoorn, 1993). We focus on those alliances that have a strategic focus and which are characterized by two-directional technology flows.[3]

We study strategic technology alliances in the microelectronics industry. We will test our hypotheses from a longitudinal perspective by examining alliance network formations. Our sample was drawn from an update of the CATI database, which covered the period 1970–2000. In the IT sector, that is computers, industrial automation, microelectronics, software and telecom, 3833 collaborative agreements were formed during this period. Strategic technology alliances in microelectronics count for 1047 alliances.

3.5 METHODOLOGY

We argued in our first proposition (P1), that alliance block formation is likely as the size of the network grows – the number of actors increases – and actors engage in multiple relationships. Alliance block formation thus results from actors developing multiple ties through local search.

To measure the size of the network, we calculated the actors active in alliance formation in the period 1970–2000. We operationalized alliance blocks by using a hierarchical clustering measure for cohesive subgroups: lambda sets at level two and four.

Concerning our second proposition (P2), we argued that when trustworthy partners are available in a densely connected social system, searching for or switching to new partners is hard to rationalize in the alliance formation process (Chung et al., 2000).

Actors rather replicate their existing ties (Gulati, 1995a, 1998; Walker et al., 1997) through local search, and they look for partners they are familiar with and with whom they share similarities in technological content in their densely connected social system.

Since our measure for alliance block membership, based on the relative frequency of ties among block members compared to non-members, does not measure cohesiveness based on in-group versus out-group strength, we need to introduce a measure that incorporates this cohesiveness in an in-group/out-group ratio similar to the one mentioned in Wasserman and Faust (1994) in order to assess the degree of replication of ties in alliance blocks. The numerator of this ratio is the number of ties that firms engage in within their group, while the denominator is the number of ties that firms form outside their core group over a certain period of time. Thus the ratio provides an indicator for in-group strength. Clearly if the ratio is higher than 1, firms are found to engage particularly in ties within their subgroup, as compared to ties outside their core block.

3.6 RESULTS

In order to test our first proposition (P1) that the formation of alliance blocks is likely as the network grows, we plotted the size of the network against the number of group members and against the number of groups in Figures 3.1 and 3.2 respectively. Figure 3.1 indicates that as the size of the network increases (that is the number of actors in the network increases), the number of group members in the network increases for lambda sets at level two and four. Figure 3.2 also shows a similar trend that the number of alliance blocks increases as the network grows, especially at level two. This seems to confirm our expectation that as alliance networks evolve into denser ones, the likelihood of the formation of alliance blocks increases.

For testing our second proposition (P2) on replication of ties within the alliance block (in-group strength) we need to use our in-group/out-group ratio as described above. Therefore we first identified densely tied group members in the lambda sets at level four. We calculated the number of in-group ties for firms at that level and divided this number by the total amount of ties that were formed in the network in that particular period. The resulting ratio indicates the percentage of ties formed in alliance blocks compared to the number of total linkages engaged in by all members of the network in that period. We focused on the periods 1980–82, 1983–85, 1986–88 and 1989–91. These ratios are shown in Table 3.1.

Table 3.1 Ratio measuring in-/out-group strength

Period	In-group/Out-group ratio	In-group/Total ties ratio
1980–82	0.6	0.2
1983–85	2.3	0.6
1986–88	3.4	0.7
1989–91	2.2	0.6

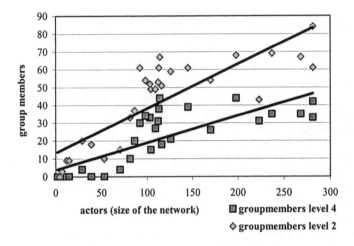

Figure 3.1 Alliance block members and network size in 1970–2000

We clearly see an increase in the in-group/out-group ratio and the in-group/total ratio for the periods 1980–82, 1983–85 and 1986–88, and a slight decrease in these ratios for the period 1989–91. This hints at the tendency of firms to replicate their ties within the subgroup as they continue to engage in new relationships over time.

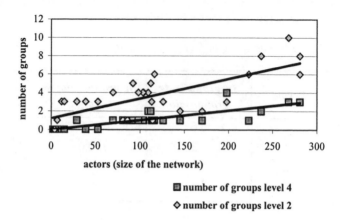

Figure 3.2 Number of alliance blocks and network size in 1970–2000

3.7 DISCUSSION AND CONCLUSION

This chapter can be seen as one of the first descriptive empirical attempts to study the process of alliance block formation from a longitudinal perspective, and points at the social mechanisms that underlie this. In order to shed more light on these subgroup formation processes, we investigated the social mechanisms that cause dynamics in inter-organizational networks and result in alliance block formation over time. Our main argument is that embeddedness is an enabling factor in the alliance network formation process, and eventually in the formation of alliance blocks, as actors invest in social capital by replicating their existing ties. We tested our main hypotheses by an empirical analysis of alliance block formation patterns in the microelectronics industry from 1970 to 2000. Our empirical results support our theoretical hypotheses as we were able to reveal the dynamics of inter-organizational networks, caused by the enabling effects of the social mechanisms we identified. We found that in an evolving alliance network, an increasing network size induces the formation of alliance blocks with firms replicating their ties within the subgroup as they engage in new relationships. This seems to support the enabling effect of embeddedness in alliance block formation, as we expected.

NOTES

1. This chapter is partly based on Duysters and Lemmens (2003).
2. Stuart and Podolny (1996) raised the concept of local search, where local search concerns initiating new R&D projects that have common technological content regarding the outcome of their prior searches.
3. See Chapter 2.

4. The Constraining Effect of Embeddedness[1]

4.1 INTRODUCTION

In the previous chapter, we addressed how social embeddedness and social capital drive the alliance network formation process in general and the block formation process in particular (Table 4.1). The social mechanisms of local search and replication of previous ties and of preferential partnering behaviour cause the network to evolve, as those mechanisms provide the enabling effect of embeddedness.

Table 4.1 The alliance network formation process

Alliance network formation		Alliance block formation	
Why do firms create ties?	*With whom do firms create ties?*	*Enabling social mechanisms*	*Constraining social mechanisms*
Strategic interdependence (exogenous)	Preferential relations through social capital (endogenous)	Social capital at group level: -Local search -Replication	Social capital at group level: -Similarity -Relational inertia

In this chapter, we will address again our first research question on the role of embeddedness in the dynamics of inter-organizational networks. However here, we describe the paralysing effect of embeddedness at the group level that is caused by constraining social mechanisms in the block formation process. We argue that the enabling effect of embeddedness during the first stages of the group formation process may turn into a paralysing effect as the block formation process progresses. We develop two propositions that address the paralysing effects of embeddedness in the worldwide microelectronics industry from 1970 to 2000. We illustrate these propositions by means of descriptive empirical evidence.

4.2 THEORETICAL BACKGROUND

As discussed in the previous chapter, the decision about with whom to partner is influenced by the network of past partnerships (Gulati and Gargiulo, 1999) and depends on the embedded relations the firm is already engaged in (Granovetter, 1985; Gulati, 1998). Because of repeated alliance formation caused by local search, frequent interaction and increased commitment in the relationship, trust and intimacy grow strong between the partners (Granovetter, 1973; Brass et al., 1998). As alliance partners have become more familiar with each other because of frequent and face-to-face contacts, 'familiarity breeds trust' (Gulati, 1995b), and greater homogeneity is expected than when those actors have fewer contacts (Wasserman and Faust, 1994: 250). 'The more tightly that individuals are tied into network, the more they are affected by group standards' (Wasserman and Faust, 1994: 250). Actors who form cohesive blocks directly influence each other through strong ties, resulting in homogeneity in attitudes, behaviour and beliefs (Wasserman and Faust, 1994: 250). Thus through the replication of their existing ties in alliance blocks, alliance block members tend to become more similar over time. This social contagion emerges when actors take up the attitudes or behaviours of others who influence them (Bovasso, 1996). Social contagion is both an individual and a group phenomenon (Burt, 1992; Bovasso, 1996). Therefore the cohesion approach suggests that similarity in attitudes stems from the proximity of actors, implying that directly linked actors will be more similar and homogeneous than indirectly linked individuals (Brass et al., 1998). This holds especially for actors that are connected by strong ties rather than weak ties (Brass et al., 1998).

As a result of the replication of ties in their group with familiar and trustworthy partners, actors may become locked-in, as they only rely on partners in their closed social system. Then, searching for or switching to partners outside of the alliance block is hard to rationalize, in particular when trustworthy partners are already available in this system. This so-called phenomenon of over-embeddedness (Uzzi, 1997), caused by the paralysing effects of embeddedness at the group level, can lead to decreasing opportunities for learning and innovation for block members involved.

Below we will address the social mechanisms that cause these constraining effects of embeddedness, by introducing our propositions.

4.3 PROPOSITIONS

Constraining embeddedness: similarity and relational inertia

Actors that are densely connected and who maintain strong ties among themselves, as in alliance blocks, are more likely to act similarly, to share information, to develop similar preferences, or to act in concert (Knoke and Kuklinski, 1982). In a similar vein, social identity theory (Gómez et al., 2000) states that similarity strengthens a self-image, as actors are attracted to similar others. Furthermore actors tend to treat those similar others more favourably than different ones (Gómez et al., 2000). Thus similarity can be the cause of attraction and the result of interaction. Scholars refer to this process as 'similarity breeds attraction' and 'interaction breeds similarity' (Brass et al., 1998).

From a technological point of view we expect that firms thus need to have some technological similarity in their technology portfolio for the replication of their ties. They require some pre-alliance technological overlap or absorptive capacity (see for example Cohen and Levinthal, 1990; Hamel, 1991; Lane and Lubatkin, 1988; Mowery et al., 1996) in order to absorb their partners' technological capabilities (Tsai, 2001). The extent to which these firms are able to learn from their partners depends on their intent. We expect that if actors have the intention to internalize their partners' technological capabilities (Hamel, 1991), instead of only accessing them, their post-alliance technological profiles will be converging and will become more similar (Mowery et al., 1996). Hence similarity can increase the block members' tendency to replicate their existing ties. Therefore we hypothesize that:

Proposition 3:
As firms replicate their existing ties within groups, their technology profiles become more similar.

Actors tend to face several endogenous constraints in the alliance network formation process. For example the resources they can devote to the search process for new partners can be limited. This means that the resources used up for forming ties with one actor can constrain them in forming ties with others (Gulati et al., 2000). Furthermore the familiarity and strong ties that have been built up through the replication of ties and the increasing similarity of firms within the alliance blocks can constrain actors in their partner choice when facing opportunities for linking up with actors of another strategic block. As a block member intends to engage in a new partnership, it can experience the implicit social pressure from its partners to replicate its ties within the group. Once firms have established links in a specific block, the formation of ties outside that block can be difficult because of conflicting interests among its partners (Nohria and Garcia-Pont, 1991). This implies that some actors in blocks are locked-in as a result of initial alliance choices and actors outside the block are locked-out. Hence there is an implicit expectation of loyalty to group members, since many alliances preclude the block members from allying with firms from competing groups (Gulati et al., 2000). As a result certain partners are not available because they are already tied to the focal firm's competitors. Another reason for locking out actors of other groups is to prevent knowledge leakage to competing groups. Finding partners outside the group is difficult as further opportunities for partnering are foreclosed by competing groups. As these groups team up with their desirable partners, they gradually become unavailable to others as the alliance formation process continues (Gomes-Casseres, 1996). Therefore some potential partners are simply excluded in the partner-selection phase.

This exogenous network phenomenon of strategic gridlock (Gomes-Casseres, 1996) forces firms to engage in local search for partners within their own strategic block (see Figure 4.1). This relational inertia makes group members rigid and cognitively locked-in (Uzzi, 1997; Gargiulo and Benassi, 2000). This cognitive lock-in effect filters the information and perspectives that reach the group members and isolates them from actors outside of the group. In this state of rigidity and over-embeddedness (Uzzi, 1997) caused by similar actors and relational inertia, alliance block members suffer from decreasing opportunities for learning and innovation. This state of over-embeddedness (Figure 4.1) is likely to put a severe strain on the block members' ability to move flexibly into other 'resource niches' or into new windows of opportunity.

In their partner choice, non-group members are restricted as well. Due to scale and scope requirements in the industry, entry barriers for non-group members rise as alliance blocks become more important (Gomes-Casseres, 1996). This tends to inhibit these non-member firms attempting to start participating in these alliance blocks and restricts their partner choice as well (Duysters and Lemmens, 2003). This leads us to the following hypothesis:

Proposition 4:
As the size of the network increases, established group members lock-out newcomers from the network.

Figure 4.1 summarizes the social mechanisms and points at the enabling and constraining effects of embeddedness they induce in alliance blocks.

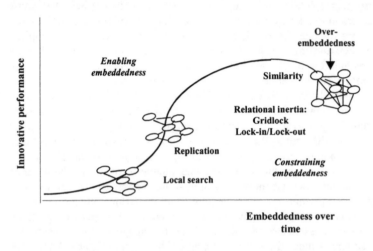

Figure 4.1 The enabling and constraining effects of embeddedness

4.4 SAMPLE AND DATA

The data on strategic alliances and characteristics of companies involved in these alliances is derived from the MERIT-CATI databank on strategic technology alliances (Duysters and Hagedoorn, 1993). We address those alliances that have a strategic focus and which are characterized by two-directional technology flows. We study strategic technology alliances in the microelectronics industry. We will test our hypotheses from a longitudinal perspective by examining alliance network formation in the microelectronics industry. Our sample of 1047 strategic alliances was drawn from an update of the CATI database, which covered the period 1970–2000.

4.5 METHODOLOGY

We argued that similarity in terms of technology profiles among block members could increase a firm's tendency to replicate its existing ties in the block. However this requires some pre-alliance technological overlap (see for example Cohen and Levinthal, 1990; Hamel, 1991; Lane and Lubatkin, 1988; Mowery et al., 1996) in order to facilitate the absorption of the partners' technological capabilities (Tsai, 2001). We expected that if actors intend to internalize their partners' technological capabilities (Hamel, 1991), their post-alliance technological profiles will be converging and will become more similar (Mowery et al., 1996).

Therefore in our third proposition (P3), we expected that as firms replicate their existing ties within groups, their technology profile would more become similar. To test this hypothesis we picked out those firms that were block members in the subsequent periods 1980–82, 1983–85, 1986–88 and 1989–91 in lambda sets at level four. These group members included: Advanced Micro Devices Inc. (AMD), International Business Machines (IBM), Intel Corp. (INTEL), Motorola Inc. (MOTOROLA), and Nippon Electric Corp. (NEC). The technology profiles were composed of the semiconductor classes in which those firms have patents (see Appendix I). We consider the similarities of these technology profiles as indicators of pre-alliance technological overlap. The technology profiles show the number of patents that a certain firm has applied for in a specific semiconductor class in a specific period (Appendix I). We expected that the technology profiles of AMD, IBM, INTEL, MOTOROLA and NEC would become more similar as they work together in an alliance block for some years in a row. Hence we assume that as a result of the learning effects associated with the strategic alliances within their group, their technology profiles as indicated by the number of applied patents for in specific semiconductor classes will become more similar over time.

Our fourth hypothesis stated that as the size of the network increases, established group members lock-out newcomers in the network. These lock-out effects result among other factors from resource constraints on forming ties with others (Gulati et al., 2000), but also from the implicit expectation of loyalty to group members, since many alliances preclude the block members from allying with firms from competing groups (Gulati et al., 2000) as argued in our theoretical section. To measure these lock-out effects, we have to investigate whether a growing number of actors in the network goes together with a relatively stable amount of group members in the network – because this indicates that these newcomers are not absorbed in groups.

4.6 RESULTS

We tested proposition three (P3) by calculating the relative differences among the technology profiles of the firms involved. We measured the amount of patents in a certain semiconductor class relative to the total amount of patent applications in all selected semiconductor classes of the firm. We calculated this ratio for all of the five block members and then subtracted those ratios in each semiconductor class from each other to find out the differences among the block members. This resulted in five outcomes per class that we summed up and subsequently divided by the number of companies to come up with the mean per class. We summed up these means per class over all semiconductor classes and this resulted in a number indicating the difference in technology profiles of the block members involved in this period.

We found in the period 1980–82 an indicator of the difference in technology profiles of the group members involved of -8; in the period 1983–85 an indicator of -3; in the period 1986–88 an indicator of -3, and in the period 1989–91 also an indicator of -3. This seems to indicate that when the firms start working together in a group, their technology profiles show some technological overlap (absorptive capacity), but are not quite similar as there are differences in their technology profiles. However after three years of replicating ties within their group, their technology profiles become more similar. After six years their technology profiles are relatively similar to those of three years before. This could be seen as an indication of decreasing learning effects and over-embeddedness, because their technology profiles do not differ compared to years before. From these results, we can conclude that by replicating ties in cohesive subgroups the technology profiles of the block members involved tend to become similar; this confirms our proposition three.

To indicate the possibility of lock-out effects (P4), Figure 4.2 points out that as the number of actors in the network increases dramatically from 1991–93 onwards, the number of group members at level two and four remained relatively stable and showed only a slight increase in this period. This may indicate that the established groups do not absorb newcomers in the network. Thus although the number of potential partners increases in the growing network, there is a possibility that they are not eligible, as they can be tied to competitors of the established group members. This would suggest that the newcomers in the network possibly form groups among themselves, as they are locked-out of the established blocks.

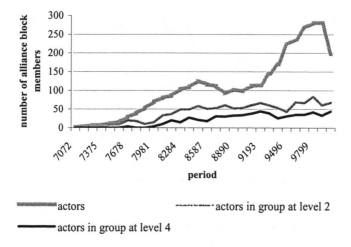

Figure 4.2 Number of alliance block members in 1970–2000

We zoom in on this phenomenon by examining Figure 4.3, which shows that from the period 1991–93 onwards, the number of groups increased at level two (the less densely tied groups). At level four, this increase started in the period 1994–96. Please note the sharp increase in the number of blocks at level two from the period 1993–95 onwards. This seems to confirm our previous suggestion that newcomers in the network possibly form groups among themselves, as they are locked-out of the established blocks. However, there is a sharp decrease starting in the period 1996–98 in the number of blocks at level two. This could be explained by the fact that these blocks are less densely tied and, as a result, are less stable and fall apart easily as the sharp decrease indicates. Altogether we find strong indications that a growing size of the network can cause lock-out effects, when the number of group members stays relatively stable. This means that we can accept proposition four.

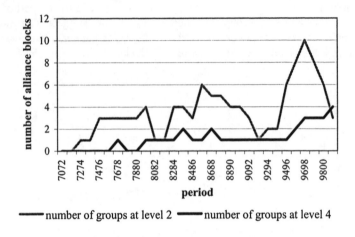

Figure 4.3 Number of alliance blocks in 1970–2000

4.7 DISCUSSION AND CONCLUSION

In the previous chapter, we were able to point at the social mechanisms that cause the dynamics of organizational networks and of alliance blocks in particular. We found that in an evolving alliance network, an increasing network size induces the formation of alliance blocks. Furthermore, we were able to show a tendency that firms replicate their ties within the subgroup as they engage in new relationships. This seemed to support the enabling effect of embeddedness in alliance block formation.

In this chapter, our empirical results supported our theoretical hypotheses, as we were able to reveal the dynamics of inter-organizational networks, caused by the constraining effects induced by the social mechanisms we identified. We showed that the enabling effect of embeddedness can turn into a paralysing effect which locks-in partners in their closed social system, and locks-out newcomers. We found empirical evidence for this in the sense that replicating ties in the group for several years increases the similarity of the technology profiles of the block members involved. This supports our arguments related to the paralysing effect of embeddedness which reduces block members' flexibility and innovative strength and which can even cause decreasing learning effects and a severe state of over-embeddedness. Moreover we found evidence that a growing size of the network can cause relational inertia and can result in lock-out effects for newcomers in the network, as our analysis showed that newcomers form blocks among themselves as they are locked out of established groups.

NOTES

1. This chapter is partly based on Duysters and Lemmens (2003).

5. Alliance Block Members: Who Are They?

5.1 INTRODUCTION

As an introductory chapter to the chapters that follow, which address the second and third research question on block membership and innovative performance, it is useful to get a better picture of the specific characteristics of block members with regard to non-alliance block members. Therefore in this chapter, we aim to get more insight in the specific attributes of alliance block members. We will investigate whether the characteristics of alliance blocks coincide with the attributes of strategic groups. The characteristics we address are: the firm's innovativeness,[1] R&D intensity and technological specialization, the firm's network position in terms of centrality and lastly its national home region and size. We will perform a discriminant analysis[2] to indicate how block members differ significantly from non-block members regarding these characteristics. To explore the formation of alliance blocks in relation to innovative performance, we address the strategic behavioural and dynamic capabilities approach to explain the strategic moves of firms in dynamic context.

5.2 THEORETICAL PERSPECTIVES ON ALLIANCE BLOCKS

We have addressed the formation and evolution of alliance blocks from a social network perspective that incorporates the social processes that drive the formation of alliances (for example Garcia-Pont and Nohria, 2002). We were able to see that alliance blocks form as the result of the social mechanisms that induce enabling and constraining effects of embeddedness, which drive the dynamic process of alliance block formation as these forces can be at work at the same time. Hence the social network perspective is crucial to explain the dynamics of the inter-organizational relations in the alliance network. Moreover, the network evolutionary perspective showed us that alliance blocks could be regarded as socio-technical systems, where inter-organizational networks co-evolve with their technologies and can result in alliance blocks.

However, if we take a broader perspective to look at the dynamic process of the formation of alliance blocks – especially in relation to innovative performance – we also have to consider some other kinds of theoretical streams of research. That is, apart from the resource-based theories, we have to incorporate those streams of research that can explain strategic moves of firms in a dynamic context. This means that these theoretical explanations have to go beyond static resource-based considerations as in the resource-based view (Dyer and Singh, 1998). The resource-based view sees the firm as a portfolio of core competencies (Prahalad and Hamel, 1990; Sakakibara, 2002) that are difficult to imitate. These competencies can also involve the firm's knowledge base. Thus to further explain alliance block formation patterns, we have to take an approach that addresses the interaction between firm resources and capabilities in a strategic context of collaboration and competition (Henderson and Mitchell, 1997; Sakakibara, 2002), where the external network is the locus of innovation and of organizational learning (Sakakibara, 2002). Here we could think of strategic behavioural theories (Kogut, 1988) and the approach of dynamic capabilities (Teece and Pisano, 1994).

The strategic behaviour approach addresses the theory that firms act to improve the competitive position *vis-à-vis* their rivals as this position influences the asset value of the firm (Kogut, 1988). In this view, collective collaborative agreements, like alliance blocks, are placed in the context of competitive rivalry to enhance market power. Thus strategic alliances are a way to adjust to changing markets and environments with the given resources a firm has. Alliances enable firms to transfer knowledge as they blur firm boundaries and hence can affect the transfer of tacit knowledge (Nonaka, 1994) and hence can replenish their knowledge bases (Mowery et al., 1996; Kogut, 1988). In this way firms engaged in multiple collaborative agreements can internalize the competencies of partners to create next-generation competencies (Hamel, 1991; Sakakabira, 2002). The performance effects of this knowledge transfer and its assimilation and internalization have been measured in the change in the number and nature of patents held (Mowery et al., 1996; Steensma and Corley, 2000). Differences in performance among competitors thus can be due to the nature of technological knowledge they possess and their ability to exploit that knowledge (Steensma and Corley, 2000).

In the dynamic capabilities approach (Teece and Pisano, 1994) the firm's ability to respond to the changing (technological) environment it is operating in is the source of competitive advantage. Multiple strategic alliances are vital instruments to develop these dynamic capabilities and to contribute to superior technological performance. These alliances complement the firm's needs in changing markets and technological environments. Strategic responsiveness is thus crucial as the time-to-market becomes critical and the pace of innovation is accelerating (Teece and Pisano, 1994).

To conclude, both the strategic behaviour and dynamic capabilities approach can explain the formation of alliance blocks, as they point at the importance of the firms' responsiveness to changing environments and characterize multiple collaborative agreements as vehicles to adjust to changing markets and environments with the given resources a firm has. In this way, engagement in multiple collaborative agreements drives the formation of alliance blocks. Hence alliance block membership improves a firm's competitive position *vis-à-vis* its rivals, as this position can influence the innovative performance of these knowledge-based firms.

5.3　STRATEGIC GROUPS VS. ALLIANCE BLOCKS

To gain more insight into the specific attributes of alliance block members, we focus on the question whether the characteristics of the alliance blocks in our sample coincide with the attributes of strategic groups in this industry.

In analysing industry structure, firms can be segmented into different strategic groups based on similarities in their strategic capabilities, for example market position, resource commitment and assets (Thomas and Venkatraman, 1988; Nohria and Garcia-Pont, 1991). Thus actors in strategic groups are similar in attributes and hence are similarly affected by environmental disturbances such as forces of globalization (Nohria and Garcia-Pont, 1991). Therefore a strategic group is akin to the ecologist's notion of a niche (Hannan and Freeman, 1977; Nohria and Garcia-Pont, 1991). Strategic groups have a distinctive source of competitive advantage, which is hard to imitate or acquire by other groups: their strategic capabilities result from the unique sequence of strategic choices; they cannot be reproduced or imitated completely (Lippman and Rumelt, 1982;

Nohria and Garcia-Pont, 1991). Thus similarity can involve specific similarities in resources or strategic capabilities of the firms involved in strategic groups. Moreover it can address similarities in certain attributes like network positions and hence point at similar relational structures.

When firms in strategic groups engage in linking with other strategic groups to adjust to changing markets and environments or to improve the competitive position *vis-à-vis* its rivals, these ties can develop into strategic blocks of connected partners within and across strategic groups. The first refers to pooling blocks composed of alliances among firms from the same strategic groups. The latter refers to complementary blocks composed of alliances among firms from different strategic groups (Nohria and Garcia-Pont, 1991). Actors in these strategic blocks regard trust as an important basis for knowledge sharing and partner selection.

As the strategic behaviour and dynamic capabilities views suggest, the firms' responsiveness to changing technological environments is crucial for knowledge-based firms. Hence multiple collaborative agreements can serve as vehicles to adjust to these conditions with the given resources these firms have. It has been argued that existing and prior relations in alliance networks strongly influence the future alliance formation process (Gulati, 1998; Hagedoorn and Duysters, 2002), as these networks facilitate the search for partners and monitor opportunistic behaviour among alliance partners (Garcia-Pont and Nohria, 2002). These streams of literature focus on the fact that firms orient their alliance behaviour to the actions of firms that have similar positions (for example central positions) or relational structures in the network through local mimetism (Garcia-Pont and Nohria, 2002). Here firms copy the behaviour of those actors they view as strategically similar, or who belong to the same strategic group. Once an alliance has been established within or across strategic groups, other members of this group will observe this behaviour and will follow as local mimetism suggests (Garcia-Pont and Nohria, 2002). As firms orient their alliance behaviour to other similar actors in this way, this dynamic can also lead to the formation of densely connected alliance blocks. Garcia-Pont and Nohria (2002) found evidence for this phenomenon in the global automobile industry. There, local mimetism was a significant driving force in alliance network formation process; little support was found for industry-wide or global mimetism. A motive for this mimitism-behaviour in alliance blocks could be to pre-empt rivals by forming alliances. This implies that there are limits to these competitive bandwagon effects (Gomes-

Casseres, 2001), as these alliance waves may lead to this situation of strategic gridlock (Gomes-Casseres, 1994, 2001; Garcia-Pont and Nohria, 2002) where the number of eligible partner diminishes as a result of overcrowding in this field (Gomes-Casseres, 2001).

From the above we can conclude that the specific characteristics of the members in alliance blocks may indeed coincide with the attributes of the members in strategic groups in this industry in cases where alliances are formed within or across strategic groups in pooling or complementary blocks (Nohria and Garcia-Pont, 1991; Garcia-Pont and Nohria, 2002). The alliance blocks that evolve can hence be the result of mimetism, where actors form alliance blocks as a result of copying the behaviour of strategically similar others within or across strategic groups (Garcia-Pont and Nohria, 2002; Nohria and Garcia-Pont, 1991).

5.4   ATTRIBUTES OF ALLIANCE BLOCK MEMBERS

In line with Duysters and Hagedoorn (2001), we perform a discriminant analysis[3] to address the specific characteristics of alliance block members in order to indicate how block members differ significantly from non-block members regarding these specific attributes or variables.

These characteristics entail the firm's propensity to patent, its R&D intensity, technological specialization, network position in terms of degree centrality (Freeman, 1979) and lastly its national home region and size. Network centrality or degree centrality is an important element of social capital as it involves benefits regarding access to resources, but also regarding information about potential partners (Gulati and Gargiulo, 1999).

Our analysis refers to a sample of 138 companies taken from the MERIT-CATI database (Duysters and Hagedoorn, 1993), which have five or more alliances during the 1980–2000 period in the international microelectronics industry (see Chapter 6).

In order see whether the group means of our grouping variable (dependent variable) differ significantly from each other, we perform a t-test.

Table 5.1 shows that our grouping variable block membership vs. non-block membership differs significantly from zero.

Table 5.2 shows the results of the discriminant analysis we performed to show how block members differ significantly from non-block members regarding a specific variable. To determine the most distinguishing variables, we start our examination of companies in the microelectronics industry with an evaluation of the Wilks' Lambda and F-values of the various variables. The Wilks' Lambda statistic is concerned with the ratio between within group variance and the total variance. A ratio that is close to one points at an equality of group means, whereas lower values are associated with large differences between the group means.

Table 5.1 One-sample t-test grouping variable

Block membership	
N	920
Mean	0.4739
Std. deviation	0.49959
Std. error mean	0.01647
t	28.773
df	919
Sign. (2-tailed)	0.000***

For each variable the F-value is calculated to test the hypothesis that all group means are equal. The results indicate that group means are not equal in the case of patent applications (0.784***), degree centrality (0.480***), Asia (0.951***), size (0.864***) and R&D intensity (0.985*). This implies a strong rejection of the hypothesis that all group means are equal for these variables and indicates that these attributes differ between alliance block members and non-alliance block members. That is, alliance block members apply for more patents than non-block members (712 vs. 161) (not corrected for firm size). Alliance block members occupy positions with a higher degree centrality[4] (10.57 vs. 1.4) than non-block members. The degree of an actor is equal to the total number of direct links of a particular actor to other actors. This implies than alliance block members have on average more direct links (10.57) than non-block members (1.4) and hence occupy more central network positions than non-block members. Furthermore our results indicate that alliance block members tend to be large firms in terms of their revenues and tend to be R&D-intensive (0.15 vs. 0.11) firms with an Asian home region.

The variables USA, Europe and specialization show high Wilks' Lambda values with insignificant results for block members and non-block members; therefore, we cannot reject the hypothesis that the group means for these variables are equal.

After we have evaluated the discriminatory power of separate variables, we continue with the overall discriminatory power of the total set of variables

We will consider the 'goodness' of a discriminant function as is reflected in various indicators presented in Table 5.3.

Table 5.2 Results of discriminant analysis

	Wilks' Lambda	F	Sig.	df1	df2	Group means: non-block members	Group means: block members
Patent applications	0.784	53.920	0.000***	1	196	161.051	712.1919
Degree centrality	0.480	212.383	0.000***	1	196	1.3977	10.5722
USA	0.990	1.955	0.164	1	196	0.7474	0.6565
Europe	0.993	1.337	0.249	1	196	0.1919	0.1313
Asia	0.951	10.041	0.002***	1	196	0.0606	0.2121
Specialization	0.999	0.233	0.630	1	196	0.4449	0.4283
Size	0.864	30.740	0.000***	1	196	7266824	19318079
R&D intensity	0.985	2.914	0.089*	1	196	0.1067	0.1491047

Note:
*Significant at the 10 % level, **Significant at the 5 % level, ***Significant at the 1 % level

The first indicator is the eigenvalue which represents the relationship of the between group and the within group sum of squares. Higher eigenvalues can be associated with a more discriminating function. In this case the function seems to have considerable discriminating power (1.292). Another important statistic is the canonical correlation (Can. Cor.) representing the proportion of total variance that is accounted for by differences among block members and non-block members. A chi-square value of 159.695 (0.000***) and a low Wilks' Lambda value of 0.436 (0.000***) imply that the hypothesis that mean scores between block members and non-block members are equal can be rejected. According to these statistics, the function has a strong discriminating power and indicates that alliance block members and non-block members do diverge with respect to a number of variables.

The effectiveness of the discriminant function is measured by classifying all cases according to their score (Table 5.4). Table 5.4 represents the classification results of the originally grouped cases. We see that 88.4 per cent of the cases are correctly classified,[5] which indicates the percentage of the cases that is correctly assigned to each of the groups (alliance block members vs. non–block members) based on the discriminant analysis.

Table 5.3 Canonical discriminant function

Function	1
Eigenvalue	1.292
% of Variance	100
Cum. %	100
Can. Cor.	0.751
Wilks' Lambda	0.436
Chi-square	159.695
df	7
***Sign. (1% level)**	0.000***

Table 5.4 Classification results

		Block lambda level 4	**Predicted group membership**		**Total**
			Non-block member (0.00)	Block member (1.00)	
Original	count	**Non–block member (0.00)**	98	1	99
		Block member (1.00)	22	77	99
	%	**Non–block member (0.00)**	99.0	1.0	100.0
		Block member (1.00)	22.2	77.8	100.0

Note: 88.4 per cent of original grouped cases are correctly classified.

5.5 DISCUSSION AND CONCLUSION

In this chapter, we tried to get a better picture of the typical characteristics of block members compared to non-alliance block members. To shed more light on the formation of alliance blocks, we addressed the strategic behavioural and dynamic capabilities approaches to explaining the strategic moves of these firms in a dynamic context.

We argued that, as a result of mimetism, actors form alliance blocks as a result of copying the behaviour of strategically similar others within or across strategic groups. Hence the specific characteristics of the members in alliance blocks may coincide with the attributes of the members in strategic groups in this industry in cases where alliances are formed within or across strategic groups in pooling or complementary blocks (Nohria and Garcia-Pont, 1991; Garcia-Pont and Nohria, 2002).

Regarding these typical attributes of alliance block members *vis-à-vis* non-block members, we addressed the firm's propensity to patent, its R&D intensity, technological specialization, network position in terms of centrality and lastly its national home region and size in a discriminant analysis. In this way we could indicate how block members differ significantly from non-block members regarding a specific variable. According to our statistics, this function has a strong discriminating power.

We found that the attribute patent applied for is a significant discriminating variable between alliance block members and non-block members, where alliance block members apply for more patents their non-group counterparts. This could indicate that a network position in terms of block membership has some implications for innovative performance (not corrected for firm size).

Another finding was that 'degree centrality' is a significant discriminating characteristic between alliance block members and non-block members; alliance block members on average had more direct links than non-block members. This can be explained by the fact that we have operationalized alliance block members by investigating their line connectivity in the group compared to the outside. This measure thus presupposes that alliance block members are hard to disconnect by the removal of ties; hence this implies that block members have a high 'degree'. A degree of an actor is equal to the total number of direct links of a particular actor to other actors. Actors that are represented by a high 'degree centrality' share the ability to access a large stock of potential information sources, which will hence contribute to their innovative performance. Furthermore central players are more visible in the network than less central players. This enhances their attractiveness to other players as it signals the firm's engagement in cooperative agreements and hence can indicate its willingness, experience and ability in strategic partnering. Thus as firms intend to enhance their own visibility and attractiveness as potential partners, they have a tendency to look for central partners (Gulati and Gargiulo, 1999). As the partners with a prominent network position are often pursued more frequently than they pursue potential partners themselves (Gulati and Gargiulo, 1999), a central position in the network is positively related to the rate of new linkage formation (Tsai, 2000).

Because of the fact that alliance block members have more direct links than non-block members, they are better-connected rivals from a competitive standpoint. Well-connected rivals may represent high-quality partners because they possess leading edge technology, have rapid access to critical information and have accumulated partnering experience (Silverman and Baum, 2002). By partnering with these better-connected rivals they may be able to turn their competitors' alliance-based competitive strengths to their own advantage. Hence tying up with well-connected rivals provides promising opportunities to learn new capabilities and to acquire advanced know-how (Silverman and Baum, 2002).

Moreover our discriminant analysis showed that size is a significant discriminating variable between alliance block members and non-block members in the sense that alliance block members are on average large firms. This could be explained by the fact that size of firms can affect the rate of R&D collaboration. There are some indications in the literature that larger companies have a higher propensity to engage in partnerships than smaller companies (Duysters and Hagedoorn, 1995; Mytelka, 1991). In the same line of reasoning, large firms could therefore also have a high propensity to engage in multiple technology alliances and hence explain their participation in alliance blocks.

Likewise, we found that the attributes of Asian home region and R&D intensity were discriminating variables between our two groups of block and non-block members, as the hypothesis of equality of group means was rejected for these characteristics. This could be explained by the fact that firms with better R&D capabilities tend to have a higher rate of participation in R&D collaborations, where past ties induce future participation (Sakakibara, 2002). A cross-sectional study on high-tech industries showed indeed that R&D-intensive firms tend to form more R&D consortia (Sakakibara, 2002). Concerning the home region, Duysters and Hagedoorn (2001) found evidence that structures and strategies of companies operating in a global environment can be still be identified with respect to their regional backgrounds. Our findings indicate that most of the block members consist of Asian firms. This could be explained by the fact that Asian firms have a long-standing tradition of collaborating with other firms in their business activities. Think of Japanese *keiretsu*, which are groups of firms with long-standing and broad-based relationships with one another in multiple fields of business (Gomes–Casseres, 1996). Likewise, Korean *chaebol* systems consist of (partially) cooperative structures, ranging from hybrid governance structures towards extremes of arm's-length, top-down contracting resembling customer–supplier relationships in vertical integration (Ahmadjian and Lincoln, 2001; Ellis and Fausten, 2002). Moreover Asian firms foster social capital in groups, as they attach value to trust-based governance in collaborative relations. This social capital represents a strong force to reproduce dense regions of ties in order to maintain and increase the value of the inherited social capital (Park and Luo, 2001).

In this chapter we pointed at the specific attributes of alliance block members. We proposed that these specific characteristics of alliance block members have an effect on their innovative performance. This chapter hence is an introduction to the next two chapters, where we explore how these alliance block members perform in terms of their innovative performance.

NOTES

1. Patent applications.
2. The major purpose of a discriminant analysis is to predict membership in two or mutually exclusive groups from a set of predictor variables.
3. For the purpose of significance testing, predictor variables should follow multivariate normal distributions; larger overall sample sizes are thus necessary to assure robustness of the method (Tabachnick and Fidell, 1996).
4. Degree centrality is measured by summing the total number of actors to which a specific player is adjacent in the matrix (a). The measure is standardized by dividing a by the maximum possible number of connections $n-1$ (n is the number of firms). In formal terms, degree centrality of firm k is equal to:

$$C_D(k) = \sum_i \frac{a_{ik}}{n-1}$$

5. Prior chance classification is 50 per cent; this means we have a high percentage of correct classifications.

6. The Innovative Performance of Block Members[1]

6.1 INTRODUCTION

The main aim of this chapter is empirically to study our second research question on block membership and innovative performance. This chapter aims to improve our understanding of how firms should position themselves in strategic alliance networks in order to maximize their innovative performance. More particularly, we empirically examine two basic technology-positioning strategies that can be pursued in terms of either alliance block membership or non-block membership. The driving forces behind the formation of these technology-driven alliance blocks are related to technology competition in groups. As technology alliances among competitors proliferate (Gomes-Casseres, 1996; Gnyawali and Madhavan, 2001), technology competition in groups becomes indispensable in the firms' technology-positioning strategies. Thus group-based technology competition affects the technology-positioning strategies at the firm level and hence the innovative performance of the firms involved, as they choose to either be an alliance block member or not.

Several authors have addressed the effectiveness of network positions by examining the effects of relational and structural embeddedness on company performance (see for example Rowley et al., 2000).

Using Coleman's (1988) closure and Burt's (1992) structural hole argument of embeddedness, we will first address the effect of closure advantages and disadvantages as well as broker advantages and disadvantages on those firms' innovative performance. We expect that a firm's innovative performance depends on its position in various network settings, that is block membership or non-block membership, reflecting closure and brokerage advantages respectively.

6.2 THEORETICAL BACKGROUND

For a long time, research on alliances has been preoccupied with the question of why and when alliances are formed (Duysters et al., 2001; Kogut and Zander, 1993; Powell and Brantley, 1992). More recently studies have dealt with the question of with whom firms are likely to form alliances (for example Gulati, 1995a; Gulati and Gargiulo, 1999), referring to how social factors, social relations and competitive tension between alliances affect the intent of creating, building and sustaining collaborative advantage through alliance formation (see for example Gulati, 1995a, 1998; Walker et al., 1997; Gulati and Gargiulo, 1999; Chung et al., 2000).

In this contribution we will address the formation of alliance blocks from a social network perspective. This perspective explains the actions of firms in terms of their position in networks of relationships (see for example Nohria, 1992; Gulati, 1998). Social networks are the embedded social relations that enclose the firms in the alliance network and indicate how these firms are connected. By investing in these relations through the replication of their prior ties, firms build up social capital (Duysters and Lemmens, 2003). Social capital relates to the investment in social relations that generates expected returns (Lin, 1999). Thus social capital refers to the prospective beneficial network of relations with external parties as well as to the resources embedded in that network that may be accessed and mobilized in purposive actions (Lin, 1999; Burt, 1992; Nahapiet and Ghosal, 1998; Chung et al., 2000). Social capital is thus dependent on history and it enables firms to rely on direct and indirect alliance experiences in partner selection (Chung et al., 2000).

Social embeddedness refers to the structure of a network of social relations and implies that the partners' ties affect the economic actions, outcomes and behaviour of firms in the network (for example Granovetter, 1992; Gulati, 1998). Firms are actually caught in a web of relations that on the one hand puts restraints on their behaviour and on the other hand can be used to their advantage. Embeddedness influences the firms' tying behaviour, because it enables preferential relations to emerge from the social capital that firms have built up through their past partnerships (Duysters and Lemmens, 2003).

Relational embeddedness focuses on the role of direct links as a mechanism for knowledge acquisition (Gulati, 1998). Structural embeddedness stresses the informational value of the position that firms occupy in the network (Gulati, 1998). Positional embeddedness refers to the impact of the positions that firms occupy in the structure of the alliance network on their decisions in the alliance formation process (Gulati and Gargiulo, 1999). Hereby, structural embeddedness (interconnectedness) creates norm creation at the network level; whereas relational embeddedness involves trust at the dyadic level (Rowley et al., 2000). We will argue that network embeddedness can be seen as an important determinant of the innovative success of companies.

Block membership can be understood as one of the strongest forms of social embeddedness. The effect of block membership on the innovative performance of companies can therefore be seen in the light of the current debate on the advantages and disadvantages of social embeddedness. In this debate on social capital (for example Rowley et al., 2000; Gargiulo and Benassi, 2000) the basic arguments stem from Burt's (1992) structural hole argument versus Coleman's (1988) closure argument. Coleman argues that being part of a dense and redundant network is advantageous since it involves trust and cooperation among its members. Hence firms engage in local search as a result of their social capital and embeddedness (Granovetter, 1985; Gulati, 1998). Because the search for partners is costly and time-consuming, firms are inclined to engage in local search for forming their subsequent ties (Duysters and Lemmens, 2003). Preferential partnering tends to reduce search costs of finding the right partners with complementary resources and eases the risk of unethical behaviour between the partners involved (Gulati and Gargiulo, 1999). Hence social capital is an important driving force in the alliance formation process (Chung et al., 2000), where the current relations of firms stem from their prior ties and they form the basis upon which the firm establishes future social ties (Gulati, 1998; Walker et al., 1997; Chung et al., 2000; Tsai, 2000). Through the replication of these ties, strategic blocks (Nohria and Garcia-Pont, 1991) or cohesive subgroups (Wasserman and Faust, 1994) of densely connected partners emerge in the strategic alliance network. These blocks or groups are characterized by highly cohesive subsets of similar firms that maintain cohesive bonds among themselves enabling them to act 'similarly, to share information, to develop similar preferences, or to act in concert' (Knoke and Kuklinski, 1982: 56).

Contrary to the above, Burt (1992) suggests that firms embedded in sparsely connected networks will enjoy brokerage advantages based on access to non-redundant information (Rowley et al., 2000). Through information access, timing, referrals and control (Burt, 1992: 62) strategic opportunities are raised as firms form bridges between densely connected and redundant parts of the network and other non-redundant parts of the network through so-called structural holes (Burt, 1992; Walker et al., 1997). Such strategies enable these firms to access knowledge or information that has a high yield. In this context, direct contacts as well as indirect contacts are found to be important. In terms of direct contacts, firms engage in local search based on social capital for extending their network. Regarding the indirect contacts, firms should look for partners that have direct links with firms they do not have strategic links with. This enables them to bridge structural holes in the network. Since current alliance networks are the basis for future alliance opportunities (Gulati, 1995a, 1995b, 1998), early participation in these networks is important. It may give those firms potentially valuable opportunities in the future (Sarkar et al., 2001). As a result, alliance-proactive firms are more likely to have the specific knowledge that is required for identifying and selecting appropriate partners in the network (Sarkar et al., 2001).

In spite of the large body of theoretical contributions, the literature is rather inconclusive about the performance effects of group membership. In order to begin filling this void we will suggest a number of hypotheses, derived from our understanding of some basic relationships between cohesive group membership and innovative performance.

6.3 HYPOTHESES

Network position and innovative performance
In most alliances, firms select partners based on prior positive experience, where they rely on their embedded relations. Partnering is thus influenced by the network of prior ties (Gulati and Gargiulo, 1999) and depends on the embedded social relations the firm is already engaged in (Granovetter, 1985; Gulati, 1998).

Members of cohesive subgroups develop strong, cohesive ties through frequent interaction. Strong ties (Granovetter, 1973) are solid, reciprocal and trustworthy relationships. They tend to create a large basis of trust and intimacy between the partners (Granovetter, 1973; Brass et al., 1998). Since trust is an important basis for knowledge sharing and joint learning, firms in these groups are expected to be more productive in their joint innovative activities. As those firms invest a substantial amount of time and energy to establish these strong relationships, changing transaction partners in the short run is not likely, since it involves substantial switching costs and implies the risk that existing relationships will dissolve (Chung et al., 2000).

Thus when trustworthy partners are readily available, searching for or switching to new partners is difficult and costly (Chung et al., 2000). Firms rather replicate their existing relations than search for new ones (Gulati, 1995a, 1998; Walker et al., 1997). As firms engage in local search, the basis of partner attractiveness will remain high and the ties between firms within blocks will even strengthen (Madhavan et al., 1998). By focusing on similar technologies in their search strategies, local search contributes to incremental innovations as firms become more competent in their technological domain and expertise (Rosenkopf and Nerkar, 2001). Furthermore, this repeated alliance formation in alliance blocks based on strong ties through local search (Duysters and Lemmens, 2003), causes the densely connected firms to behave similarly and to develop similar preferences (Knoke and Kuklinski, 1982). Similarity can in turn encourage interaction and can hence form the basis of attraction. Scholars refer to this process as 'interaction breeds similarity' and 'similarity breeds attraction' (Brass et al., 1998). Similarity – from a technology point of view – means that firms have some similarity in their technology profile, which is required to assimilate and understand the technology they have access to when replicating their ties (Duysters and Lemmens, 2003). Thus some pre-alliance technological overlap (see for example Cohen and Levinthal, 1990; Hamel, 1991; Lane and Lubatkin, 1998; Mowery et al., 1996) or similar past R&D activities (Rosenkopf and Nerkar, 2001) is required to absorb their partners' technological capabilities. If those firms intend to internalize their partners' technological capabilities (Hamel, 1991) and learn from them instead of only accessing these capabilities, their post-alliance technological profiles will be converging and will become more similar (Mowery et al., 1996; Duysters and Lemmens, 2003).

In the context of strong ties and familiarity, joint innovative activities and the sharing of knowledge in alliance blocks are expected to generate higher innovative performance than when firms follow an individual innovation strategy outside cohesive subgroups. Hence,

Hypothesis 5:
Members of cohesive subgroups are more innovative than non-member firms.

In many cases, the enabling effect of embeddedness in alliance formation that is based on replication of preferential relations can turn into a paralysing effect. Then, firms can become locked-in to their alliance blocks, as they only rely on partners in their own closed social system (Duysters and Lemmens, 2003). Over time those firms may even start to suffer from relational and technological 'over-embeddedness' (Uzzi, 1997) caused by relational inertia and the increasing similarity of firms within the alliance blocks.

Block members can be constrained in their partner choice when linking up with firms of another alliance block. For example there may be limits to the resources they can devote to the search process for new partners. This implies that the allocation of resources for the formation of ties with one firm, can constrain them in forming ties with others (Gulati et al., 2000). Additionally, block members can experience implicit social pressure from their partners to replicate their ties within the alliance block (Duysters and Lemmens, 2003). Once firms have established ties in a specific alliance block, the formation of links outside that block can be constrained, because of conflicting interests among its partners (Nohria and Garcia-Pont, 1991). Furthermore, there is this implicit expectation of loyalty to group members, which prevents block members from allying with firms from competing alliance blocks (Gulati et al., 2000). As a result certain partners cannot be selected, because they have ties to the block members' competitors. In this way firms in blocks can be locked-in resulting from initial alliance choices, and firms outside the alliance block can be locked-out. To prevent knowledge leakage to competing groups, firms can lock-out actors of other alliance groups (Duysters and Lemmens, 2003).

As competing groups can foreclose further partnering opportunities, finding partners outside a core alliance group is difficult (Duysters and Lemmens, 2003). As those alliance groups team up with their desirable partners, these partners gradually become unavailable to others as the alliance formation process proceeds (Gomes-Casseres, 1996). Hence potential partners are simply not available in partner selection. This phenomenon of strategic gridlock (Gomes-Casseres, 1996) forces block members to engage in local search for partners within their own alliance block and hence makes them relationally inert. Thus too much focus on developing technological competencies through local search can lead firms to develop core rigidities (Leonard-Barton, 1995). Furthermore they can cause firms to fall into competency traps (Levitt and March, 1988). This inertia and rigidity among group members makes them cognitively locked-in (Uzzi, 1997; Gargiulo and Benassi, 2000). The cognitive lock-in effect isolates block members from firms outside of the alliance group, as it filters the information and perspectives that reach the block members from the outside. In this state of rigidity, collective blindness and over-embeddedness (Uzzi, 1997), alliance block members suffer from decreasing opportunities for learning and innovation (Duysters and Lemmens, 2003), as they are restrained to take advantage of new opportunities and resource niches (see Chapter 4). Therefore in terms of learning we expect that, over time, over-embeddedness and similarity lead to decreasing opportunities for learning and innovation (see Figure 6.1). Thus:

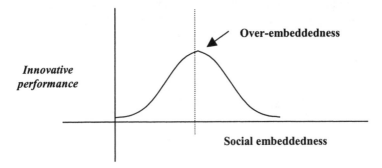

Figure 6.1 Over-embeddedness

Hypothesis 6:
There is a curvilinear (inverted-U shaped) relationship between alliance block membership and innovative performance.

To prevent developing these core rigidities in groups, firms need to move away from local search and need to reposition themselves by allying with firms that give them access to different and new information by, for example bridging structural holes in the network (Burt, 1992; Walker et al., 1997). However group members do not always have this option, as social pressure and loyalty to alliance block members can hamper these group members' actions. This can result in an implicit tension within the group, as alliance block members have to outweigh the advantage of moving into a new technology that can enhance their innovative performance versus the disadvantage of the reputation effect that results from leaving the group. This tension can lead to dissolution of these alliance blocks eventually.

6.4 SAMPLE AND DATA

Our analysis refers to a group of 138 companies taken from the MERIT-CATI database (Duysters and Hagedoorn, 1993), which have five or more alliances during the 1980–2000 period we studied and are dominantly present in the international microelectronics industry. By calculating the degree centralities in UCINET (Borgatti et al., 1999, 2002) over the period 1980–2000, we found 138 companies who had five alliances or more in this period[2] (Appendix II). Because we did not have patent data available for three companies, we worked with a sample of 135 companies (see Appendix II), of which 68 are American, 35 are European and 32 Asian.

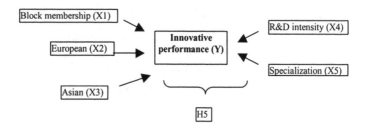

Figure 6.2 The expected linear model 1

6.5 METHODOLOGY

To measure the effect of alliance block membership on innovative performance (H5) (see equation 6.1 and Figure 6.2), we performed a standard ordinary least square regression.[3] The residuals of the dependent variable innovative performance, measured by the patent intensity of firms, that is the number of patents applied for divided by firm size (revenues) are normally distributed after a logarithmic transformation (see Appendix III). This means we can perform linear regression analysis (Tabachnik and Fidell, 1996). This is confirmed by the further analysis of scatterplots of the residuals (see Appendix IV), where the normal probability plot of the standardized residuals approximates the diagonal that indicates a normal distribution. We control for the firm's technological specialization, R&D intensity and home region (see Figure 6.2).

$$Y = C + \beta_1 X_1 + \beta_2 X_2 + \beta_3 X_3 + \beta_4 X_4 + \beta_5 X_5 \qquad (6.1)$$

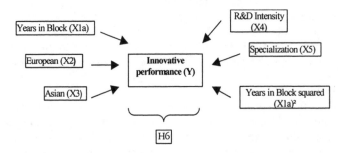

Figure 6.3 The expected curvilinear relation model 2

To measure the curvilinear (inverse U-shaped) relation between alliance block membership and innovative performance (H6) (see Figure 6.3), we also use an OLS regression model, although we add a higher order quadratic term to it (Equation 6.2). If we suspect that we are dealing with a curvilinear relation, thus if we think there is a bend in the curve describing the relation between our dependent and independent variable, we fit a quadratic model to the data:

$$Y = C + \beta_1 X_{1a} + \beta_2 X_2 + \beta_3 X_3 + \beta_4 X_4 + \beta_5 X_5 + \beta_6 \left(X_{1a} \right)^2 \quad (6.2)$$

Subsequently, we will perform a curvilinear regression analysis which we will carry out hierachically, using stepwise regression. We will control for the firm's technological specialization, R&D intensity and home region. We start with the linear model (see Table 6.3) and then add progressively the higher order term (in this case a second order, quadratic term), that is the squared variable year block. If this term produces a significant change in R^2, over and above all lower-order terms (that is linear), we can state that the type of curvilinear relation it represents does describe the data.

6.6 VARIABLES AND MEASURES

Dependent variable

The dependent variable innovative performance, we measure by the patent intensity (10log) of firms, which is the number of patents applied for divided by firm size (revenues) in the period 1983–2000.

We summed up the patent applications per firm per year for our sample of 135 companies in the 15 microelectronics patent classes in periods of three years. Thus period 1 is 1983–85, period 2 is 1984–86, period 3 is 1985–87 and so on. Then we divided the number of patents applied for by the revenues, in order to correct for firm size.

Independent variable

Our first explanatory variable is alliance block membership or cohesive subgroup membership. We measure block membership by calculating lambda sets at the clustering level four. This means we operationalize alliance block membership by investigating line connectivity in the group compared to outside. Alliance block members are thus hard to disconnect by the removal of ties. We assign a dummy 1 for block members and a dummy 0 for non-block membership. In the theoretical section of this chapter we addressed the structural hole argument (Burt, 1992) versus Coleman's (1988) closure argument to describe the advantages and disadvantages of social embeddedness and its strongest form that we characterized as alliance block membership. However in operationalizing this construct, we do not make a distinction between the several roles alliance block members can occupy, like for example core or periphery players in the alliance block (Everett and Borgatti, 1999a, 1999b). The same holds for non-alliance block members, as we do not distinguish in this group between brokers occupying a structural hole position or non-alliance block members occupying a peripheral position in the network. In other words we put all non-alliance block members' roles in one group (non-alliance block membership) and we lump all alliance block members together in the alliance block membership group.

To measure the curvilinear relation (H7) indicating over-embeddedness, we also need a variable measuring the duration of block membership, where we count the years of unbroken presence in an alliance block: years in block *(X1a)* (see Appendix V). We derive this variable from our variable block membership.

Control variables

As discussed in the methodology chapter we rely on three control variables: the national home region of companies, that is the United States, Europe and Asia, and the R&D intensity (ratio of R&D expenditures to total revenues). Additionally we incorporate the control variable of technological specialization, where we divide the patents applied for in our 15 semiconductor classes by the total amount of patents applied for per company per period. Since we correct for firm size (revenues) in the dependent variable, we do not incorporate size as an additional control variable in this analysis.

6.7 RESULTS

To measure the effect of alliance block membership on innovative performance (H5), we apply an OLS regression. Table 6.1 lists the correlation coefficients between the variables. In order to detect possible multicollinearity in our data we performed a number of multicollinearity tests (see Appendix VI). Table 6.2 presents the results of the multiple regression analysis for our microelectronics sample and the means for the variables in the analysis (model 1). According to the F value (63.398***) and the adjusted R^2 value (0.619), model 1 is significant. We find a positive relation for block membership and innovative performance (0.120***). This confirms our reasoning that prior cohesiveness based on frequency of ties within the subgroup compared to the outside (Alba, 1973) matters for block members' lagged innovative performance. As those subgroups develop solid, reciprocal and trustworthy relationships, they are more productive in their joint innovative efforts. This confirms our hypothesis five (H5).

As opposed to hypothesis five (H5), where we tested whether block membership positively influences innovative performance, we test in hypothesis six (H6) the effect of the duration of being a block member and innovative performance. We argue that alliance block members become over-embedded after some years in a block by replication of ties (see also P3) and this will hence influence their innovative performance. Thus here our independent variable becomes years in block, instead of block membership. In other words, in hypothesis six we expect that there is a inversed U-shape (curvilinear) relation between years of block membership and innovative performance.

To test the curvilinear relation (H6) between innovative performance and the years of unbroken presence in an alliance block, we use stepwise regression, where we enter the variables successively (Table 6.3), including the squared term. We find an F-change of 10.898*** and an R^2 change of 0.014, if we add the squared term year block to our linear model. This change in F is significant 0.001*** and implies that indeed a curvilinear model describes the data.

Table 6.1 Pearson correlation coefficients of models 1 and 2

		1	1a	2	3	4	5	6
Model 1	1. Block Membership	1.000						
	2. US	-0.127 **		1.000				
	3. Europe	-0.05		-0.658***	1.000			
	4. Asia	0.218***		-0.631***	-0.170***	1.000		
	5. R&D Intensity	0.123**		0.144 **	-0.027	-0.159**	1.000	
	6. Spec.	-0.035		0.310***	-0.243***	-0.155**	0.277***	1.000
Model 2	1a. Years in Block		1					
	2. US		0.081*	1				
	3. Europe		0.017	-0.616***	1			
	4. Asia		-0.117**	-0.647***	-0.202***	1		
	5. R&D Intensity		0.336***	0.149***	-0.007	-0.179***	1	
	6. Spec.		0.127**	0.289***	-0.182***	-0.183***	0.276***	1

Note: *Significant at the 10 % level, **Significant at the 5 % level, ***Significant at the 1 % level

Table 6.2 Regression estimates of models 1 and 2

	Model 1 (block membership and innovative performance) N = 193		Model 2 (over-embeddedness and innovative performance) N = 291	
Constant	(-62.934)***		(-75.798)***	
	Beta	**Mean**	**Beta**	**Mean**
Block membership	0.120*** (2.597)	0.5078		
Years in block			0.141*** (3.630)	2.57
Europe dummy	-0.208*** (-4.423)	0.150	-0.275*** (-7.188)	0.1615
Asia dummy	-0.061 (-1.279)	0.139	-0.081** (-2.106)	0.1753
USA dummy		0.709		0.663
R&D intensity	0.201*** (4.245)	0.129	0.160*** (3.973)	0.1087
Specialization	0.618*** (12.730)	0.441	0.579*** (14.737)	0.396
Adj. R²	0.619		0.615	
St. Err.	0.44890		0.55602	
F	63.398***		93.761***	
DW	0.956		0.827	

Note:
*Significant at the 10 % level, **Significant at the 5 % level, ***Significant at the 1 % level;
Numbers in parentheses are t-statistics

However in this curvilinear model we also found that there is a high correlation between the years in block variable and the variable years in block squared (0.969***); adding this squared term thus increases the multicollinearity between the variable years in block and years in block squared, as the VIF values rise to around 17.[4] This indicates that adding the squared term means we are losing precision of the estimates.

Model 2 in Table 6.2 shows that the linear term 'years in block' is significant with a positive beta of 0.141***. This indicates that the shape of the line fitting our data has a strong linear upward sloping component to it.

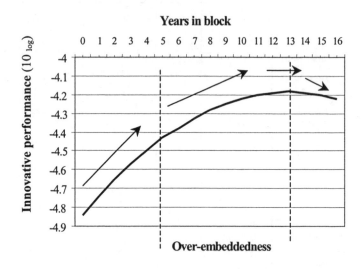

Figure 6.4 The curvilinear relation between innovative performance and block membership model 2

A visual examination of our data (see Figure 6.4) by fitting a power polynomial mean[5] to our data, hints again at a curvilinear relationship: a upward sloping line ending in an inversed U-shape[6] that describes the relation between innovative performance (the 10log of patent intensity) and the years of unbroken presence in a block. We see an almost linear increase in the means of patent intensity until year five. Pursuing a block membership strategy for more than five years up to 13 years implies decreasing returns to scale (Figure 6.4), indicating a first sign of over-embeddedness. Then, over-embeddedness truly sets in from year 13. Moreover Figure 6.4 shows that the patent intensity starts to decline for actors that stay in the alliance block for one to three years longer from that point on.

This finding supports our argument that pursuing a block membership strategy for some years has a curvilinear impact on the innovation output of these firms, which could point at over-embeddedness; confirming hypothesis six (H6).

Table 6.3 Change statistics from linear to curvilinear model

Model	R Square Change	F Change	df1	df2	Sign. F Change	Durbin-Watson
1	0.498	286.719	1	298	0.000***	
2	0.057	36.548	1	288	0.000***	
3	0.043	30.583	1	287	0.000***	0.792
4	0.019	13.847	1	286	0.000***	
5	0.014	10.898	1	285	0.001***	
6	0.010	7.777	1	284	0.006***	

Notes:
*Significant at the 10 % level, **Significant at the 5 % level, ***Significant at the 1 % level

1. predictors: Constant, Spec.
2. predictors: Constant, Spec., Europe
3. predictors: Constant, Spec., Europe, R&D intensity
4. predictors: Constant, Spec., Europe, R&D intensity, Years in block
5. predictors: Constant, Spec., Europe, R&D intensity, Years in block, Years in block squared
6. predictors: Constant, Spec., Europe, R&D intensity, Years in block, Years in block squared, USA

Concerning our control variables (corrected for firm size), we find for both models 1 and 2 that specialization in microelectronics and R&D intensity positively affects the firms' patent intensity in this sector. Additionally we find a negative relation for European and Asian firms[7] and their patent intensity. American firms positively relate to innovative performance.

6.8 DISCUSSION AND CONCLUSION

This chapter aimed to improve our understanding of how firms should position themselves in order to maximize their innovative performance. Our empirical results supported our theoretical hypotheses.

For hypothesis five (H5) we found that a block membership strategy positively influences innovative performance.[8] This can be due to the fact that members of cohesive subgroups develop strong ties characterized by solid, reciprocal and trustworthy relationships. Since trust is often referred to as a precondition for knowledge sharing, firms are expected to be more productive in their joint innovative efforts. Through the use of strong ties, firms take advantage of the network externalities in their alliance block, as these solid relationships are a means to transfer tacit knowledge in this learning environment founded on trust-based governance.

Concerning hypothesis six (H6), we found indications that block members become over-embedded as a result of rigidity and hence become less innovative, as this state of over-embeddedness is likely to put a severe strain on a their ability to move flexibly into other 'resource niches' or into new windows of opportunity. We found, as expected, a curvilinear relation between the years of continuous presence in a block and its effect on innovative performance. We found decreasing returns to scale for firms pursuing a block member strategy for five years or more. This indicates the first signs of over-embeddedness. Pursuing a block membership strategy for 13 years or more is not benefical for innovative performance, as patent intensity starts to decline. Earlier findings in Chapter four (P3) are in line with these findings. There we used the similarity in technology profiles as an indicator of over-embeddedness. We found that the block members' technology profiles started to converge after six years, which could point at first signs of over-embeddedness. Then as block members become more rigid, because they are exposed to redundant information for some years, this could lead to decreasing learning effects and result in over-embeddedness. Hence we can imagine that this similarity has a negative effect on the block members' innovative performance in the long run.

Because we did not find evidence that both alliance block members and non-block members differ from each other regarding their technological specialization,[9] this could imply that the motive behind the local search for partners is not necessarily based on a specific technology to contribute to their innovative performance. Alternatively this local search for partners could be based on the social mechanisms and trust-based governance that bind these alliance blocks together and hence provide a basis for joint innovative efforts.

To conclude, in order to maximize innovative performance in an inter-organizational network, a network position as a block member seems to be a rewarding strategy. However we have reason to believe that there is a limit to the positive effect of this strategy on innovative performance, as we found evidence for over-embeddedness which can seriously hamper a block member's innovative performance.

In this chapter, we investigated certain network positioning strategies and their effect on innovative performance. However, we did not take into account the state of the technological environment. Therefore in the next chapter we will address the question of how firms should position themselves in strategic alliance networks under various technological conditions, in order to maximize their innovative performance. With technological conditions we refer to technological events that are either structure-reinforcing or structure-loosening (Madhavan et al., 1998), caused by incremental technological developments or disruptive technologies (for example Bower and Christensen, 1995), respectively.

NOTES

1. This chapter is partly based on Duysters et al. (2003).
2. The degree of an actor is equal to the total number of direct links of a particular actor to other actors.
3. In this analysis we did not incorporate both the variables alliance block membership and degree centrality in one analysis, as they point at overlapping roles in the network and hence show signs of multicollinearity.
4. Tolerances are found to be near 0 and the variance proportions are near 1 (cut-off 0.5).
5. Regression prediction line based on mean with 95 per cent confidence interval.
6. The fact that the beta (-0.566***) for the added quadratic term is negative (not depicted here), indicates that the curvilinear relation it represents has the form of an inversed U-shape, as we expected.
7. This result is significant only in model 2.
8. In the empirical analysis we did not make a distinction between the roles alliance block members could occupy in the alliance block membership group. Likewise we did not operationalize broker positions (occupying structural holes) specifically in the non-alliance block membership group.
9. See the discriminant analysis in Chapter 5.

7. Technological Change and Performance of Block Members[1]

7.1 INTRODUCTION

In the previous chapter we empirically examined two basic technology-positioning strategies that can be pursued in terms of either alliance block membership or non-block membership. We found evidence for the fact that an alliance block membership strategy positively affects innovative performance, as these firms take advantage of the network externalities in their block through the use of strong ties. However we also found indications that there is a limit to the positive effect of this strategy on innovative performance, as we found evidence for over-embeddedness which can seriously hamper a block member's innovative performance.

We have explained that multiple strategic alliances in alliance blocks are vital instruments for developing dynamic capabilities and to improve technological performance, as these alliances act as vehicles to complement the firm's needs in changing markets and technological environments.

Therefore the main aim of this chapter is to improve our understanding of how technology-based firms should position themselves in strategic alliance networks in order to maximize their innovative performance, taking into account the turbulence of the technological environment. We expect that a firm's innovative performance depends on its position in various network settings (block membership or non-block membership) and is affected by the nature of technological change (cumulative vs. disruptive).

Hence we will empirically test some basic hypotheses on the moderating effect of technological turbulence on network position and innovative performance. This chapter intends to answer our third research question.

7.2 THEORETICAL BACKGROUND

Many authors have argued that strong ties are particularly effective under conditions of relative stability, whereas weak ties are particularly geared towards dynamic industry environments (for example Rowley et al., 2000; Uzzi, 1997; Larson, 1992). Others (for example Hagedoorn and Duysters, 2002) found that under conditions of turbulence, a learning strategy employing many, seemingly redundant, alliances might be more effective to increase firm performance than a maximizing strategy that is geared towards bridging structural holes.

Although we share the notion of some of these authors that 'the degree of uncertainty and required rate of innovation in the environment influence the appropriate network configurations' (Rowley et al., 2000: 370), we argue that these findings are above all contingent on the stage of a network's evolution. With the latter we refer to the evolution of a network that is either structure-reinforcing or structure-loosening (Madhavan et al., 1998), caused by incremental technological developments or disruptive technologies (for example Bower and Christensen, 1995), respectively.

We expect that a firm's innovative performance depends on its position in various network settings (block membership or non-block membership) and is affected by the nature of technological change (cumulative vs. disruptive). We will derive some basic hypotheses on the effect of block membership on innovative performance under various technological conditions.

7.3 HYPOTHESES

Network position, technological change and innovative performance

In contrast to the general conception of new life cycles born out of market needs (Sherwin and Isenson, 1967; Utterback, 1974) high-technology industries are typical examples of markets created by radical technological innovations (Mueller and Tilton, 1969; Tushman and Anderson, 1986). At these early stages, there is often substantial uncertainty about the technological feasibility of an innovation and its potential market size.

Ultimately those technologies that are most successful – in both technological terms and in meeting customer demands – will accumulate a critical mass and may set a new technological regime.

The emergence of a new technological regime leads to a substitution of radical technological development by more focused incremental cumulative improvements along a specific technological trajectory (Dosi, 1988; Duysters, 1996). Incremental technological improvements are structure-reinforcing since they enhance and extend the underlying sustaining technology, and thus reinforce the existing status quo (Tushman and Anderson, 1986; Madhavan et al., 1998) and bases of competition. Such technologies are also accumulative and competence enhancing (Tushman and Anderson, 1986) and support the way the industry is functioning.

The establishment of a technological regime does not only lower technological uncertainty. Due to its adaptation in terms of new market standards, market uncertainty is also reduced considerably (Dosi, 1988). From that point onward, cumulative improvements in technology are becoming more important than radical innovation. Since the industry is characterized by accumulative technological improvements, which are structure-reinforcing, incremental innovation occurs through the interaction of many firms (Tushman and Anderson, 1986). By focusing on similar technologies in their local search strategies, incremental innovations thrive, as firms become more competent in their technological domain and expertise (Rosenkopf and Nerkar, 2001). This might lead to a situation in which cohesive subgroups thrive, whereas firms that are particularly effective in bridging structural holes (Burt, 1992) are less effective. Under these conditions, when innovation depends on a series of interdependent innovations, independent companies will have a hard time coordinating and tying these innovations together (Chesbrough and Teece, 1996). Hence we expect firms to integrate these innovations by engaging in strategic block formation. Through strategic block formation, firms within blocks can enhance and extend the underlying sustaining technology (Tushman and Anderson, 1986). In this way block members exploit their existing capabilities by linking up with firms in their own technology cluster to improve their innovative performance. Hence:

Hypothesis 7:
In a situation of structure reinforcing cumulative technological change, alliance block members are more innovative than their non-block counterparts.

After a period of technological progress and considerable market growth, most industries undergo a phase of more moderate technological and market development. Saturation of demand levels sales growth towards zero, whereas technological progress seems to approach its natural limits. Faced with problems of advancing current technologies, firms need to invest an increasing amount of resources in R&D to make significant new progress. In order to speed up stagnating technological progress, firms should broaden their focus in search for alternative technologies (Duysters, 1996). These search processes may eventually lead to new technological regimes or to the establishment of a new technological paradigm. Substitute technologies may offer better perspectives and may be able to trigger off new technological paths. These radical and disruptive technological innovations do often drastically alter the price/performance ratio of high-technology products and act as forces of creative destruction, which threaten incumbent industry leaders and open up opportunities for new firms (Duysters and De Man, 2003). Under these structure-loosening conditions it might be sensible for any organization to shift its attention towards the new technological paradigm (Duysters, 1996). This competence-destroying discontinuity (Tushman and Anderson, 1986; Madhavan et al., 1998) alters the way the industry functions and can radically change the bases of competition in an industry. The shift in regime also reshuffles both the current bases of attractiveness and the existing ties of firms in blocks and may thus result in an out-block orientation in partner selection.

When technological change is radical and disruptive (Bower and Christensen, 1995), the impetus for cohesive group members to move into this technology is not very high. The reputation effects within the group are not offset by the potential rewards that can be found in engaging with these innovations. Furthermore most cohesive subgroup members are characterized by strong inertial forces, which prevent them from entering into more innovative new relationships. The implicit expectation of loyalty to alliance block members prevents them from allying with firms from competing groups (Gulati et al., 2000) as they experience the implicit social pressure from their partners to replicate ties within the group. The close ties and commitment the block members share in their technological community can develop into a collective blindness to the outside, which makes them vulnerable to disruptive changes in the environment. Group pressures might even lead to situations in which incumbents tend rather to increase investments in the old technology than to switch to the new technological regime (Foster, 1986). The inability of these cohesive subgroup members to explore new technologies paves the way for non-alliance block members to take advantage of the new technologies. Non-alliance group members may have created a radar function of alliances in order to scan the most promising technologies. They can expect high rewards for bringing a technologically new product to the market. Thus:

Hypothesis 8:
Under conditions of structure-loosening disruptive technological change, non-alliance block members have a higher innovation rate than their alliance block member counterparts.

7.4 SAMPLE AND DATA

The sample and data we use for testing the hypotheses in this chapter are the same as used in Chapter 6. Our analysis refers to the same group of 135 microelectronic companies taken from the MERIT-CATI database (Duysters and Hagedoorn, 1993) having more than five alliances during the 1980–2000 period we studied.

7.5 METHODOLOGY

To measure the effect of alliance block membership on innovative performance under various technological conditions, i.e. cumulative or disruptive technological change, we performed an OLS regression.[2] We controlled for the firms' technological specialization, R&D intensity and home region. We introduce an interaction effect in our model (Figure 7.1). In our hypothesis seven (H7), we expect that under cumulative technological change a block membership position is beneficial for innovative performance. In hypothesis eight (H8) we expect that under disruptive technological conditions a non-block membership position is desirable. This means that for both hypotheses the outcome of the interaction effect *(K1)*, that is, the multiplication of our variables years in block *(X1a)* and technological change *(Z1)* has to have a positive value, so that it will positively affect innovative performance in the regression equation. Therefore, our regression equation looks like this:

$$Y = C + \beta_1 \left(Z_1 X_{1a} \right) + \beta_2 X_2 + \beta_3 X_3 + \beta_4 X_4 + \beta_5 X_5 \qquad (7.1)$$

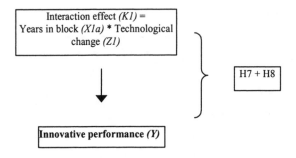

Figure 7.1 The expected interaction effect model 3

7.6 VARIABLES AND MEASURES

Dependent variable
As a measure for our dependent variable innovative performance we took the same measure as in our previous chapter: the patent intensity (10log) in the microelectronics sector during the period 1983–2000.

Independent variable
In this chapter, we need a variable measuring the duration of block membership, where we count the years one is unbroken present in an alliance block: years in block *(X1a)*. We thus derive this variable from our variable block membership (Appendix V) as used in our previous chapter.

Moderator variable
To test our hypotheses seven and eight, we introduce the moderator variable interaction effect *(K1)*. This moderator variable is required to address the interaction between the years one is pursuing an alliance block membership strategy *(X1a)* and the nature of technological change *(Z1)* and its effect on innovative performance *(Y)*.

Technological change measures turbulence in the technological environment. This variable compares the relative differences in technology profiles at the industry level in relation to the base year of the previous period. Higher numbers indicate turbulence or disruptive technological change; lower numbers of technological change indicate cumulative technological development. We measure technological change by calculating the total amounts of patents applied for in the various microelectronics technology classes per year.

Table 7.1 Dissimilarity matrix

	1: Case 1	2: Case 2	3: Case 3	4: Case 4	5: Case 5	6: Case 6
1: Case 1 (1980—82)		0.102716	0.416366	0.617444	0.735678	1
2: Case 2 (1983—85)	0.102716		0.234918	0.410455	0.532443	0.835402
3: Case 3 (1986—88)	0.416366	0.234918		0	0.38374	0.755032
4: Case 4 (1989—91)	0.617444	0.410455	0		0.269788	0.627099
5: Case 5 (1992—94)	0.735678	0.532443	0.38374	0.269788		0.066187
6: Case 6 (1995—97)	1	0.835402	0.755032	0.627099	0.066187	

Then we calculate the percentage of patents applied for in a specific microelectronics class as a percentage of the total amount of patents applied for in microelectronics in a certain period. Subsequently we performed a hierarchical clustering analysis, which resulted in a dissimilarity matrix (Table 7.1) based on Euclidean distances. This matrix shows the distances between relative differences in technology profiles at the industry level in certain periods.

By comparing the periods in relation to each other (for example 1983–85 in relation to 1980–82), we can derive the variable technological change *(Z1)*. Thus the technological change from 1983–85 in relation to 1980–82 is 0.102716[3] (Table 7.1 and Table 7.2). The technological change in this period thus affects the block members in the subsequent periods: 1981–83, 1982–84 and 1983–85 and hence their innovative performance in 1984–86, 1985–87 and 1986–88 incorporating the time lag of three years.

Table 7.2 Variable technological change

Variable Technological Change (Z1)	
case 2 vs. case 1	0.102716
case 3 vs. case 2	0.234918
case 4 vs. case 3	0
case 5 vs. case 4	0.269788
case 6 vs. case 5	0.066187

However, we see in our dissimilarity matrix and the variable technological change *(Z1)* (see Table 7.2) that a small value indicates that the technological change has been relatively small with respect to the previous period, which points at cumulative technological change. This small value multiplied by the number of years in a block should have a large positive effect on our dependent variable. However, this is not the case here, then the effect of a high value of technological change indicating disruptive change multiplied by the years in a block generates a higher positive effect on our dependent variable; this is counter-intuitive to our hypotheses.

Table 7.3 Z-score normalizations

	Years in Block (X1a)	*z-Years in Block (X1a norm.)*
Non-Block member	0	-0.35351
	1	0.09577
	2	0.54505
	3	0.99433
	4	1.44361
	5	1.89290
	6	2.34218
	7	2.79146
Block member	8	3.24074
	9	3.69002
	10	4.13930
	11	4.58858
	12	5.03787
	13	5.48715
	14	5.93643
	15	6.38571
	16	6.83499
	Change (Z1)	*z-Change (Z1 norm.)*
	0.00000	1.36186
Cumulative	0.06619	0.69272
	0.10272	0.32342
	0.13471	0.00000
Disruptive	0.23492	-1.01313
	0.26979	-1.36566

We solve this problem by normalizing the technological change values using z-score normalization.[4] The resulting normalized range of values will have a mean of 0 and a standard deviation of 1. We model our interaction effect by applying z-score normalization to our variables technological change *(Z1)* and years in block *(X1a)*, that both make up the interaction effect. Therefore, we normalize our variable technological change *(Z1)* by taking its technological change values minus the mean of all technological change values divided by the standard deviation (see Table 7.3). We do the same for our variable years in block *(X1a)*. As a result of z-score normalization, our disruptive change values have a negative value[5] and our non-block membership value (0 years in block) also has a negative value (Table 7.3). This is required, since we expect that non-block members are innovative under disruptive change (for example -0.35351 * -1.36566) and alliance block members are especially innovative under cumulative technological change (for example 0.54505 * 0.69272). Thus, by using z-score normalization, we are able to model our interaction effect in such a way, that the outcome of the interaction effect *(K1)* for both cases is positive; in this way the interaction effect will positively affect innovative performance in the regression equation. The interaction effect is thus operationalized as the z-scores of the technological change values multiplied by the z-scores of years one is unbroken present in an alliance block (equation 7.2):

$$Y = C + \beta_1 \left(Z_{1norm.} X_{1anorm.} \right) + \beta_2 X_2 + \beta_3 X_3 + \beta_4 X_4 + \beta_5 X_5 \qquad (7.2)$$

Control variables
We use the same three control variables as in the previous chapter: the home region of companies, the R&D intensity and technological specialization.

7.7 RESULTS

To measure both the interaction effects of cumulative technological change and alliance block membership on innovative performance (H7) and disruptive technological change and non-block membership on innovative performance (H8), we applied an OLS regression, meeting the same conditions for normality as in the previous chapter.

After a visual examination of our data (see Figure 7.2), we have reason to believe that the relation between patent intensity and the interaction effect is not linear. By fitting a polynomial mean[6] to our data, we find a curvilinear U-shape relation between innovative performance (the 10log of patent intensity) and the moderator variable: the interaction effect.

Table 7.4 Change statistics interaction effect

Model	R Square Change	F Change	df1	df2	Sign. F Change	Durbin-Watson
1	0.498	286.719	1	289	0.000***	
2	0.057	36.548	1	288	0.000***	0.809
3	0.048	34.647	1	287	0.000***	
4	0.021	15.860	1	286	0.000***	

Notes:
*Significant at the 10 % level, **Significant at the 5 % level, ***Significant at the 1 % level

1. predictors: Constant, Specialization
2. predictors: Constant, Specialization, Europe
3. predictors: Constant, Specialization, Europe, Squared Interaction Effect
4. predictors: Constant, Specialization, Europe, Squared Interaction Effect, R&D ratio

In order to test this curvilinear relation, we have to add the squared term of our variable interaction effect to our regression model. We use stepwise regression, where we enter the variables successively, including the squared term to investigate whether this squared term contributes significantly to the model.

In Table 7.4 we find an F-change of 34.647*** and an R^2 change of 0.048, if we add the squared interaction effect to our model.[7] This change in F is significant (0.000***), and implies that indeed a curvilinear model in the form of a U-shape describes the data.

Table 7.5 shows the correlation coefficients between the variables in the curvilinear relation (model 3). In order to detect possible multicollinearity in our data we performed a number of multicollinearity tests (see Appendix VI). Table 7.6 presents the results of the multiple regression analysis for our sample including the means for the variables in the analysis. We see that model 3 is significant according to the F value (96.160***) and adjusted R^2 value (0.621). Furthermore, the squared interaction effect (0.164***)[8] positively influences our dependent variable innovative performance. This points at the earlier found U-shaped form of our curve.

This U-shaped curve (Figure 7.2) implies that the interaction effects at the right-hand side (values from 3 to 6) and the interaction effects and the left-hand side (-5 to -7.5) score high on innovative performance.

When we zoom in on the interaction effects at the right hand side of Figure 7.3, these values from 3 to 6 (see Table 7.3) involve alliance block members who are positioned in a block for six years or more under very cumulative technological change (for example z-score years in block * z-score technological change = 2.34218 *1.36186 = 3.1890).

Network position and technological change

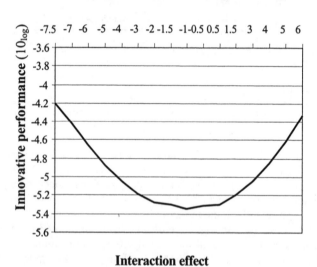

Interaction effect

Figure 7.2 The curvilinear relation between innovative performance and the interaction effect model 3

This implies that occupying a block membership position for several years in a very cumulative technological environment is beneficial for innovative performance. This finding confirms hypothesis seven (H7).

Likewise, the interaction effects at the left-hand side of Figure 7.3, the values from -5 to -7.5 (see Table 7.3) involve alliance block members who are positioned in a block for ten years or more under very disruptive technological change (for example z-score years in block * z-score technological change = 4.1393 * -1.36566 = -5.653). This implies that contrary to our expectations, occupying a block membership position for a long period of time in a very turbulent and disruptive technological environment is beneficial for innovative performance. This finding rejects hypothesis eight (H8).

Table 7.5 Pearson correlation coefficients model 3

Model 3	1	2	3	4	5	6
1. US	1					
2. Europe	-0.616***	1				
3. Asia	-0.647***	-0.202***	1			
4. R&D Intensity	0.149***	-0.007	-0.179***	1		
5. Spec.	0.289***	-0.182***	-0.183***	0.276***	1	
6. Inter-action effect[2]	0.096*	0.060	-0.177***	0.341***	0.120**	1

Note:
*Significant at the 10 % level, **Significant at the 5 % level, ***Significant at the 1 % level

The interactions effects at the middle of the U-shape (see Figure 7.3) score low on innovative performance. These interaction effects include the non-block members that operate under disruptive change (for example z-score years in block * z-score technological change = -0.35351 * -1.36566 = 0.358). This finding is contrary to our expectations and again rejects hypothesis eight (H8). Furthermore the interaction effects at the middle of the U-shape (see Figure 7.3) also include non-block members, which score low on innovative performance under cumulative technological change, as we expected. This follows from an investigation of the values of the interaction effects in the middle of Figure 7.3 (for example z-score years in block * z-score technological change = -0.35351 * 1.36186 = -0.481). This again confirms hypothesis seven (H7).

Concerning our control variables, we find a negative relation between European and Asian firms and their patent intensity (Table 7.6) and the opposite for American firms. Additionally we find that technological specialization and R&D intensity positively affects innovative performance. This is similar to our findings in the previous chapter where we did not include the interaction effect.

Table 7.6 Regression estimates interaction effect model 3

	Model 3 (Interaction effect) N = 291	
Constant	(-76.680)***	
	Beta	Mean
Europe dummy	-0.280*** (-7.377)	0.162
Asia dummy	-0.071* (-1.841)	0.175
R&D intensity	0.152*** (3.821)	0.109
Specialization	0.580*** (14.888)	0.396
Interaction effect squared	0.164*** (4.233)	5.624
Adj. R²	0.621	
St. Err.	0.55165	
F	96.160***	
DW	0.818	

Note:
*Significant at the 10 % level, **Significant at the 5 % level, ***Significant at the 1 % level
Numbers in parentheses are t-statistics

7.8. DISCUSSION AND CONCLUSION

The main aim of this chapter was to investigate the role of turbulence of the technological environment that moderates the relation between network position and its effect on innovative performance. Our basic assumption was that a firm's innovative performance depends on its position in various network settings (block membership or non-block membership) and is affected by the nature of technological change (cumulative vs. disruptive).

In line with our expectations, we found that pursuing a block membership strategy under cumulative (H7) change is valuable for innovative performance.

Under these technological conditions, joint innovative efforts through strategic block formation ameliorate the coordination and the linking of these innovations, together enhancing and extending the underlying sustaining technology.

Network position and technological change

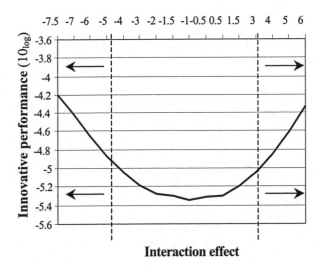

Figure 7.3 The curvilinear relation model 3

Relying on other players in the group gives better chances for innovative renewal as a result of spillover effects, which enables block members to tap into each other's knowledge base. In this way block members exploit and deepen their existing capabilities by linking up with firms in their own technology cluster to improve their innovative performance.

Non-block members have a low innovative performance under cumulative technological change in comparison to alliance block members, as we expected (H7).

In the previous chapter where we did not take the technological environment into account in our model, we found decreasing learning effects in terms of over-embeddedness for alliance block members after six years of block membership. In this chapter we find evidence that, after the same period of being a block member, pursuing a block membership strategy contributes positively to innovative performance under cumulative technological changes. This implies that this point of over-embeddedness manifests itself in a later phase in case of incremental technological changes based on interdependencies among members. These technological conditions foster the virtues of closure and block membership, which have a positive impact on innovative performance.

However, contrary to our expectations we did not find empirical evidence that non-alliance block members perform well in a disruptive and turbulent technological environment (H8). Therefore we have to reject hypothesis eight (H8). Actually our findings indicate that non-block members perform badly under disruptive change. We expected that disruptive change would be favourable for non-block members, as they can manoeuvre flexibly into new technological opportunities without facing the inertia that block members would have to face in that situation. Since we have only investigated a network position that can involve either pursuing a block membership or non-block membership strategy, and we did not differentiate between the several roles non-block members can occupy in the network like periphery players or brokers occupying structural holes, we can not judge whether non-block members occupying a broker position could perform better than alliance block members under disruptive change. According to the network literature, broker or structural hole positions in the network are valuable as a structural hole connects several (closed) parts of the network (Burt, 1992). Actors occupying this role have access to new information, which means they can explore the technological environment for new opportunities.

An additional striking finding concerning hypothesis eight was that alliance block members perform well under disruptive change. This means that anticipating this disruptive change in a group works out better than doing this individually. Apparently, the trust-based governance we find in alliance blocks is not only beneficial under conditions of cumulative change, as we found, it is also beneficial under turbulent technological change. An explanation for this can be that the trust and embeddedness found in the block can help block members to deal with the novel technologies related to the turbulent environment. Block membership enables members to deepen, exploit and even stretch their existing knowledge base. This ability to stretch their technological know-how is an important competence that is required to discover the technological and commercial benefits of new technologies and disruptive changes (Cohen, Levinthal, 1990). In the same line of reasoning, alliance block membership can help the actors to manage the ambiguity related to the disruptive change, as these multiple alliances are a mechanism to keep options open (Gomes-Casseres, 2001).

These findings also have implications regarding our arguments on over-embeddedness, in the sense that our hypotheses suggest that in case of disruptive change the constraining effects of embeddedness will surface earlier for block members, as rigidity and inertial effects would be effectuated earlier in this situation. Again, we could not find evidence for this, because our results indicate that alliance block members also perform well under disruptive change.

From these findings we can conclude that elaborating on cumulative technological innovations and struggling with newness caused by disruptive technological changes is easier to handle in a group than alone. We found that alliance block members are fairly R&D-intensive;[9] this finding could also point at the fact that alliance block members have a sophisticated technological base. This gives them the possibility to stretch their technological knowledge in related high-tech industries, thus surpassing only the microelectronics field. Through this ability they are better able to overcome crisis in turbulent technological environments.

NOTES

1. This chapter is partly based on: Duysters et al. (2003).
2. The residuals of the dependent variable are normally distributed.
3. Since this matrix does not give a value for the base year 1980–82, we calculated this value by taking the mean of our variable technological change *(Z1)* values: 0.13471.
4. Z-score normalization uses the mean and standard deviation of the values for each attribute to standardize the values for that attribute.
5. After multiplication with −1.
6. Regression prediction line based on mean with 95 per cent confidence interval.
7. The linear term interaction effect is excluded in the stepwise regression analysis as this term is not significant (0.717) and will therefore not be included in our final model 3.

8. The fact that the beta (0.164***) for the added quadratic term is positive indicates that the curvilinear relation it represents has the form of an U-shape.
9. See the discriminant analysis in Chapter 5.

8. Discussion and Conclusions

8.1 INTRODUCTION

In this chapter we will answer our research questions by discussing the findings of the hypotheses that we have tested. We will draw conclusions in terms of whether we have reached our research objective for this research question, and address the limitations and suggestions for further research on this issue. Finally, we will conclude this chapter by giving some final remarks.

The focus of this study was conceptually to build on the academic work concerning social structure and its effect on alliance formation patterns in social networks (see for example Gulati, 1995a, 1998; Granovetter, 1992; Gulati and Gargiulo, 1999; Walker et al. 1997) by integrating these streams of research. Furthermore we aimed empirically to study network evolution in inter-organizational networks in high-tech industries from a longitudinal perspective. More particularly, we intended to further the theoretical development of the concept of alliance blocks or 'technology driven constellations' (Gomes-Casseres, 1996) by taking a social network perspective (for example Nohria, 1992; Gulati, 1998). Therefore we aimed at identifying the social mechanisms that cause enabling and constraining effects of embeddedness and network dynamics in technology alliance blocks. The dynamics of inter-alliance networks are increasingly driven by these social mechanisms that follow from embeddedness and investing in social capital (Gulati, 1995a; Gulati and Gargiulo, 1999). These social mechanisms cause the inter-organizational networks to self-generate, self-transform and self-reinforce in alliance blocks. In this study we tried empirically to identify the social component that is gaining importance over the technological aspect as a driving force in the network evolutionary process. Then, firms increasingly look for trustworthy and preferential relations through replication of their existing ties to improve their innovative performance.

Moreover this study intended to contribute conceptually as well as empirically to the current body of literature on embeddedness and network positioning strategies of firms in alliance networks – in alliance blocks in particular – and their effect on innovative performance (for example Rowley et al. 2000; Gulati et al., 2000; Gargiulo and Benassi, 2000; Coleman, 1988; Burt, 1992) under changing technological conditions (Madhavan et al., 1998; Bower and Christensen, 1995).

Therefore we have conducted a quantitative research project based on an extensive literature study and on an empirical analysis of the CATI alliance database in the microelectronics industry. As this research project is theory-oriented and deductive in nature, we have derived hypotheses from the academic literature in this field that we have tested empirically.

8.2   THE DUAL ROLE OF SOCIAL STRUCTURE IN THE ALLIANCE NETWORK FORMATION PROCESS

Since the focus of our study was to build conceptually as well as empirically on the theory of social structure and its effect on alliance formation patterns in social networks, we are interested in empirically identifying the social mechanisms that induce those network dynamics as they can cause enabling and constraining effects of embeddedness. Apart from the network-enabling effect of embeddedness (for example Granovetter, 1992; Gulati, 1998) in alliance block formation – where embeddedness is a driving factor in the network evolutionary process – the academic literature has given considerably less attention to the constraining effects of embeddedness in the decision with whom to partner. In many cases the enabling effect of embeddedness in alliance formation that is based on preferential relations can turn into a paralysing effect as actors become locked-in, as they only rely on partners in their own closed social system. Therefore we have addressed the following research question:

1. *What is the role of embeddedness and its social mechanisms in the dynamics of inter-organizational networks and alliance blocks specifically?*

Regarding the enabling role of embeddedness that follows from the social mechanisms local search and replication of existing ties, which cause the dynamics of inter-organizational networks and alliance blocks specifically, we proposed that when the size of the network increases, alliance block formation becomes more likely (P1). This hypothesis was based on our expectation that firms replicate preferential relations based on their social capital (Lin, 1999) and their embeddedness. In this way, the network self-generates and reproduces over time. Being embedded in a densely connected network as a result of a high amount of social capital, makes engagement in subsequent ties more likely (Walker et al. 1997) and transforms these relations in strong ties (Granovetter, 1973). As a result, actors develop multiple cohesive ties through local search, which increases the possibility that alliance blocks are formed.

To test this proposition, we plotted the size of the network against the number of group members and against the number of groups. We found that as the size of the network increases, the number of group members in the network increases as well as the number of groups. This descriptive empirical evidence seems to confirm our expectation that, as alliance networks evolve into denser ones, the likelihood of the formation of alliance blocks increases. To zoom in on this phenomenon that we have established, we are interested in why firms tend to replicate their relations in alliance blocks. In the literature we find that transaction costs play a role in the sense that changing transaction partners in the short run is not likely. This involves significant switching costs and implies a risk that existing relationships will dissolve (Chung et al., 2000). Furthermore firms tend to replicate their existing partnerships because of the danger of reputation effects. This fear deters firms in a web of relations from behaving opportunistically against each other, and it increases the stability and longevity of alliance formation in their closed system.

To test our expectation of in-group strength, we propose that when firms look for new partners, they replicate their ties within the subgroup (P2). For testing this proposition on replication of ties within the alliance block (in-group strength), we have developed an in-group to out-group ties ratio and in-group to total ties ratio. Descriptive empirical evidence shows that these ratios increase in the period of observation, which hints at the tendency of firms to replicate their ties within the subgroup as they continue to engage in new relationships over time.

To summarize, we found descriptive empirical evidence that indicates that embeddedness and the social mechanisms that follow from it (for example local search and replication of ties) cause dynamics in the inter-organizational network and result in alliance block formation over time. These social mechanisms thus have an enabling effect in the evolution of inter-organizational networks and cause the network to self-generate and self-transform, which may eventually lead to the formation of alliance blocks.

The constraining role of embeddedness, which follows from the social mechanism of similarity that can induce attraction and result in interaction (Brass et al., 1998), causes the dynamics of inter-organizational networks and alliance blocks in particular. We hypothesized that replication of ties among actors within alliance blocks leads to similar technology profiles (P3). This hypothesis was based on our expectation that firms in alliance blocks directly influence each other through strong ties, resulting in homogeneity in attitudes, behaviour and beliefs (Wasserman and Faust, 1994). Thus through the replication of their existing ties in alliance blocks, alliance block members tend to become more similar over time, which can lead to decreasing learning effects and even cause a state of over-embeddedness. To test this proposition, we have calculated the relative differences among the technology profiles of the firms involved that work together in alliance blocks. We found that when firms start working together in a group, their technology profiles show some pre-alliance technological overlap in terms of absorptive capacity. However after several years of replicating ties within their group, their technology profiles become more and more similar. After six years their technology profiles are relatively similar to those three years before. From these descriptive empirical results, our expectation is confirmed in that replication of ties in cohesive subgroups leads to similarity in the technology profiles of the block members involved. This could be seen as a first indication of decreasing learning effects and over-embeddedness.

Concerning the constraining role of embeddedness that follows from the social mechanism of relational inertia that causes the dynamics of inter-organizational networks and alliance blocks, we hypothesized that when the size of the network increases, established group members lock-out newcomers from the network (P4). This hypothesis was based on our expectation that as a result of the replication of ties in their group with familiar and trustworthy partners, actors may become locked-in, as they only rely on partners in their closed social system. Searching for or switching to partners outside of the alliance block is hard to rationalize, in particular when trustworthy partners are already available in this system. To test this proposition on these lock-out effects, we have investigated whether a growing number of actors in the network goes together with a relatively stable amount of group members in the network; because this indicates that these newcomers are not absorbed in groups. We found for our period of observation that the number of actors in the network increases dramatically, whereas the number of group members remained relatively stable and showed only a slight increase in this period. This may indicate that the established groups do not absorb newcomers in the network. Thus although the number of potential partners increases in the growing network, there is a possibility that they are not eligible, as they can be tied to competitors of the established group members. This would suggest that the newcomers in the network possibly form groups among themselves, as they are locked-out of the established blocks. We find descriptive evidence for this as we find a sharp increase in the number of blocks.

To summarize, we were able to reveal the dynamics of inter-organizational networks, caused by the constraining effects of embeddedness induced by the social mechanisms we identified (that is similarity and relational inertia).

Conclusion
Altogether, we have reached our research objective to conceptually as well as empirically build on the theory regarding social structure and its effect on alliance formation patterns in social networks (see for example Gulati, 1995a, 1998; Granovetter, 1992; Gulati and Gargiulo, 1999; Walker et al., 1997) we integrated these streams of research and empirically studied the social mechanisms that cause dynamics in the evolution of inter-organizational networks from a longitudinal perspective. We found descriptive empirical evidence for the dual role of social structure in the alliance network formation process and alliance block formation in particular. We found proof for the enabling effect of embeddedness that can turn into a paralysing effect which locks-in partners in their closed social system, and locks-out newcomers. This phenomenon of so-called over-embeddedness, caused by the paralysing effects of embeddedness at the group level, can lead to decreasing opportunities for learning and innovation for block members involved. The social mechanisms of local search and replication of ties thus have an enabling as well as a constraining effect in the evolution of inter-organizational network as they eventually cause similarity and relational inertia. These social mechanisms thus result in a social structure, which cause the network

to self-generate, self-transform and self-reinforce, which may eventually induce the formation of alliance blocks.

Limitations and further research
The descriptive empirical analysis on the dual role of social structure in the alliance network formation process and alliance block formation in particular has its limitations in terms of the degree to which we can generalize its outcomes; in particular, concerning the periods of study. To establish the general trends on the growing size of the network and the likelihood of the formation of alliance blocks, and the possibility of lock out effects, we have tested in the period 1970–2000. For the establishment of in-group strength caused by replication of ties and for pointing at the increasing similarity in technology profiles caused by this, we have focused on the period 1980–91. Furthermore our empirical evidence is only descriptive in nature and covers only the microelectronics industry.

Therefore future research might provide further insight into the enabling and constraining effects of embeddedness and its relation to the dynamics of network evolution, through more in-depth empirical research. This in-depth empirical research should incorporate also other high-tech industries, where the timeframe of the object of study could be extended.

8.3 BLOCK MEMBERSHIP AND ITS EFFECT ON INNOVATIVE PERFORMANCE

We have investigated the formation and evolution of alliance blocks from a social network perspective that incorporates the social structure that drives the formation of alliances (for example Garcia-Pont and Nohria, 2002). We were able to see that alliance blocks form as the result of the social mechanisms that induce enabling and constraining effects of embeddedness, which drive the dynamic process of alliance block formation as these forces can be at work at the same time. Hence the social network perspective is crucial to explain the dynamics of the inter-organizational relations in the alliance network. The network evolutionary perspective showed us that alliance blocks are socio-technical systems, where inter-organizational networks co-evolve with their technologies and can result in alliance blocks.

To investigate the dynamic process of the formation of alliance blocks in a broader perspective – especially in relation to innovative performance – we have considered theoretical streams of research that go beyond static resource-based considerations as in the resource-based view (Dyer and Singh, 1998). Therefore in this study we have incorporated those streams of research that can explain strategic moves of firms in a dynamic context: strategic behavioural theories (Kogut, 1988) and the approach of dynamic capabilities (Teece and Pisano, 1994). To explore the formation of alliance blocks in relation to innovative performance, we address this strategic behavioural and dynamic capabilities approach to explain the strategic moves of firms in dynamic context. This means we have placed collective collaborative agreements, like alliance blocks, in the context of competitive rivalry to enhance market power. These multiple strategic alliances are vital instruments for developing dynamic capabilities and to contribute to superior technological performance, as these alliances act as vehicles to complement the firm's needs in changing markets and technological environments. Thus both the strategic behaviour and dynamic capabilities approach explain the formation of alliance blocks, as they point at the importance of the firms' responsiveness to changing environments which is crucial as the time-to-market becomes critical and the pace of innovation is accelerating (Teece and Pisano, 1994). In this way, engagement in multiple collaborative agreements drives the formation of alliance blocks, as pursuing an alliance block membership strategy improves a firm's competitive position *vis-à-vis* its rivals and hence influences the innovative performance of these technology-based firms.

To gain more insight in the specific characteristics of alliance block members, we have empirically studied the specific attributes that distinguish them from non-alliance block members and which can influence their innovative capability. By means of a discriminant analysis, we were able to reveal that alliance block members – compared to their non-block counterparts – apply for more patents and hold more central positions in the network. Furthermore alliance block members are large firms in terms of their revenues. They undertake more R&D-intensive research and are inclined to originate from an Asian background in comparison to non-alliance block members. We proposed that these specific characteristics of alliance block members could have an effect on their innovative performance.

Since the academic literature is rather inconclusive about the performance effects of multiple collaborative agreements in general and of alliance block membership in particular, as it is a debated issue, this study intended to contribute conceptually as well as empirically to the performance effects of firms in alliance networks (for example Rowley et al., 2000; Gulati et al., 2000; Gargiulo and Benassi, 2000; Coleman, 1988; Burt, 1992). Since alliance block membership is the strongest form of social embeddedness (for example Burt, 1992; Coleman, 1988; Rowley et al., 2000; Gargiulo and Benassi, 2000), the effect of block membership on the innovative performance of companies can therefore be seen in the light of the current debate on the advantages and disadvantages of social embeddedness (for example Burt, 1992; Coleman, 1988; Rowley et al., 2000; Gargiulo and Benassi, 2000). We intended to contribute empirical evidence to this debated issue of embeddedness and its effect on innovative performance. Therefore, we formulated the following research question:

2. What is the effect of embeddedness – alliance block membership in particular – on innovative performance?

We hypothesized that members of cohesive subgroups are more innovative than non-member firms (H5). This hypothesis was based on our expectation that in an alliance block, joint innovative activities and the sharing of knowledge are expected to generate higher innovative performance as a result of strong ties and familiarity, than when firms follow an individual innovation strategy outside alliance blocks.

To test this proposition we performed a standard ordinary least square regression with innovative performance (patent intensity) as a dependent variable and alliance block membership strategy as the independent variable, and we included some control variables. We found a positive relation between block membership and innovative performance. This confirms our reasoning that prior cohesiveness based on frequency of ties within the subgroup compared to the outside (Alba, 1973) matters for block members' innovative performance. As those subgroups develop solid, reciprocal and trustworthy relationships, they are more productive in their joint innovative efforts.

However as we have seen in our analyses there is a limit to the advantage of operating in a closed social system such as in an alliance block. Over time those alliance block members may start to suffer from relational and technological over-embeddedness, caused by relational inertia and the increasing technological similarity of firms within the alliance blocks. We hypothesized that there is a curvilinear (inverted-U shaped) relationship between alliance block membership and innovative performance (H6).

We test this curvilinear relation by adding the squared independent variable years in block to our regression model, and find an almost linear increase in the relation between the duration of following an alliance block membership strategy and innovative performance. However, from five years in the block until 13 years in the block, we evidenced decreasing returns to scale, which indicate first signs of over-embeddedness. Additionally we found that over-embeddedness truly sets in if actors follow a block membership strategy for more than 13 years.

Conclusion

In order to maximize innovative performance in an inter-organizational network, a network position as a block member seems to be a rewarding strategy. However due to the constraining effect of embeddedness, we have reason to believe that there is a limit to the positive effect of this strategy on innovative performance, as we found evidence for over-embeddedness which can seriously hamper a block member's innovative performance.

Altogether we have reached our research objective to contribute empirically as well as conceptually to the current body of literature on network positioning strategies of firms in alliance networks and in alliance blocks in particular. We found empirical evidence to establish the performance effects of firms embedded in alliance blocks through multilateral collaborative technology agreements. This means we were able to build on the state-of-the-art theoretical streams of research to show the effects of embeddedness – alliance block membership in particular – on innovative performance (for example Rowley et al., 2000; Gulati et al., 2000; Gargiulo and Benassi, 2000; Coleman, 1988; Burt, 1992). This implies that we were able to empirically build on the theory of Coleman's closure advantages regarding network position and performance effects.

Limitations and further research

In our analyses we looked at the performance effects of firms pursuing either a block membership strategy or a non-block membership strategy. However in this way we treated the innovative power of all block members as equal, and hence overlooked that some blocks might be more innovative than others, due to some specific players in the block that could leverage the overall innovative capacity of the group. Future research could therefore focus on the differences in the innovative capabilities of alliance blocks and should therefore focus on the specific attributes the most innovative groups have. It would be helpful to gain insight in this by investigating a larger number of industrial sectors to increase our further understanding the effect of alliance block membership and innovative performance.

8.4 BLOCK MEMBERSHIP AND ITS EFFECT ON INNOVATIVE PERFORMANCE UNDER TECHNOLOGICAL CHANGE

We have explained that multiple strategic alliances in alliance blocks are important instruments for developing dynamic capabilities and to improve technological performance. To include the moderating effect of the degree of technological change on network positioning strategies and innovative performance of technology-based firms, we addressed the following research question:

3. *How does technological change – that is either disruptive or incremental – mediate the relationship between block membership and innovative performance?*

To answer this question we initially hypothesized that in a situation of structure-reinforcing cumulative technological change, alliance block members are more innovative than their non-alliance block counterparts (H7). This hypothesis was based on our expectation that under these conditions of cumulative technological change, alliance block members have the advantage of linking these innovations together in order to enhance and extend the underlying sustaining technology (Tushman and Anderson, 1986). In this way block members exploit their existing capabilities, as they are connected to firms in their own technology cluster to improve their innovative performance. To test this hypothesis, we performed an OLS regression, where we included a moderator variable pointing at an interaction effect including the normalized technological change values multiplied by the normalized values of the years of unbroken presence in an alliance block. We found a curvilinear relation between this interaction effect and its effect on innovative performance. This U-shaped relation pointed at the fact that, in line with our expectations, pursuing a block membership strategy under cumulative change is valuable for innovative performance. Relying on other players in the group gives alliance block members better chances for innovative renewal as a result of spillover effects in their technology cluster to improve their innovative performance as they can tap into each other's knowledge base.

This supports our expectation that these specific technological conditions foster the virtues of closure and block membership as they have a positive impact on innovative activity.

Concerning the moderating effect of disruptive change on alliance block membership and innovative performance, we hypothesized that under conditions of structure-loosening disruptive technological change, non-alliance block members have a higher innovation rate than their alliance block member counterparts (H8). This hypothesis was based on our expectation that non-alliance members do not suffer from group-based pressures and relational inertia that increases 'groupthink' and makes them blind to what is happening outside the alliance block. We expected that disruptive change thus would be favourable for non-alliance block members, as they can manoeuver flexibly into new technological opportunities without facing the inertia that block members would have to face in that situation. We found, contrary to this expectation, that non-alliance block members perform worse than block members in a disruptive and turbulent technological environment.

An additional striking finding was that block members not only perform well under cumulative change, but also under disruptive change. This means that dealing with disruptive technological change in a group of alliances is more effective than doing this on an individual basis. Apparently the trust-based governance we find in alliance blocks is not only beneficial under conditions of cumulative change, as we found, it is also beneficial under turbulent technological changes. Alliance block membership can help the actors to manage the ambiguity related to the disruptive change, as these multiple alliances are a mechanism to keep options open within the group of partners (Gomes-Casseres, 2001). These findings also have implications regarding our arguments about over-embeddedness, in the sense that our hypotheses suggest in cases of disruptive change the constraining effects of embeddedness will surface earlier for block members, as rigidity and inertial effects would be effectuated earlier in this situation. Again we could not find evidence for this, because our results indicated that alliance block members perform well under disruptive change.

Conclusion

From these findings we can conclude that elaborating on cumulative technological innovations and struggling with newness caused by disruptive technological changes is easier to handle in a group of alliances than on an individual basis. We found earlier that alliance block members are fairly R&D-intensive; this finding could also point at the fact that alliance block members have a sophisticated technological base. This gives them the possibility to stretch their technological knowledge, which enables them to discover the technological and commercial benefits of new technologies and disruptive changes (Cohen and Levinthal, 1990). Through this ability they are better able to overcome crisis in turbulent technological environments.

Because we did not find evidence that alliance block members and non-block members differ from each other regarding their technological specialization, this could imply that the motive behind the local search for partners is not necessarily based on a specific technology to contribute to their innovative performance. Alternatively this local search for partners could be based on the social mechanisms and trust-based governance that bind these alliance blocks together and hence provide a basis for joint innovative efforts.

Altogether we have reached our research objective in the sense that we were able to contribute conceptually as well as empirically to the academic work in this field (for example Madhavan et al., 1998). We found that the degree of turbulence in the environment has an impact on the firm's network configurations and signals the direction of the network's evolution. This implies that if a certain network position in combination with a certain degree of turbulence in the technological environment, does not yield innovative renewal, a firm has to change its network positioning strategy to find a network position that does provide a satisfying yield under these circumstances. These strategic moves can also induce a network dynamic and cause the network to evolve as it changes the basis of competitiveness in the industry.

Limitations and further research

Although this study on the innovative performance of alliance block members is limited to one industrial sector, which means we have to be careful to generalize its outcomes, the examined sector is a large and strategic sector. The latter has enabled us to systematically focus on and explore the basic questions related to network strategy and innovative performance, without the disturbances one could possibly encounter in multi-industry designs. Nevertheless future research might provide further insight to the relation between embeddedness and innovative performance through more in-depth qualitative empirical research incorporating other high-tech industries.

In our study we have investigated a network position that can involve pursuing an alliance block membership or non-block membership strategy. In the empirical analysis we did take into account that non-block members can occupy several roles in the network. Future research could incorporate for example broker positions into the analysis in order to get a clearer picture of the performance of the group of non-alliance block members. Firms bridging structural holes might increase their learning capabilities and might become fairly innovative as they get access to new information and can act as a broker in the network with as little redundancy as possible (for example Granovetter, 1973; Burt, 1992; Hagedoorn and Duysters, 2002). Including broker positions in the research implies that we could truly investigate whether a block membership position, which benefits from the virtues of network externalities in a block, would weigh up against a network position outside of an alliance block that bridges a structural hole.

More generally, future research could incorporate more diverse network positioning strategies within alliance blocks as well as outside of alliance blocks in relation to innovative performance. Therefore it would be interesting to look at the relation between the several roles that non-block members can occupy in the network as mentioned above. Apart from the broker position that we expect to perform well, we could also look at more peripheral players and their innovative performance. Here we would expect that more peripheral players in loosely-coupled networks outside of alliance blocks would be less innovative. Concerning block membership strategies, one could incorporate the more diverse roles block members can occupy. Here one has to think in terms of core periphery structures (Everett and Borgatti, 1999a, 1999b) in alliance blocks, where true core players or true periphery players could perform differently in terms of their innovativeness.

Furthermore, future research could further elaborate on how the network dynamics that are caused by the effects of social embeddedness could affect group-based competition specifically, as this issue was beyond the scope of this research. Likewise further research could be directed at social factors that relate to the internal organization of alliance groups, as this issue was beyond the scope of this research. Here we refer to operational aspects related to the management and coordination of groups.

8.5 FINAL REMARKS

This volume has made a conceptual as well as an empirical contribution in its research field as it has studied how social structure causes the alliance network to evolve. The social mechanisms we established can induce enabling and constraining embeddedness in technology alliance blocks and cause the network to self-generate, self-transform and self-reinforce. This study also established that the dynamics that cause the evolution of alliance networks, and alliance blocks in particular, are also affected by the degree of technological change in the environment. As we have seen this degree of technological change has an effect on the effectiveness of a firm's network position in terms of its innovative performance. This implies that if a certain network position does not yield innovative renewal under specific technological conditions, a firm has to change this network positioning strategy in such a way that it does provide a satisfying yield under these circumstances. These strategic moves can also induce a network dynamic and cause the network to evolve. In the same line of reasoning, turbulence in the technological environment can also induce a network dynamic, in the sense that it can arouse an alliance wave (Gomes-Casseres, 2001) and can lead to strengthening of alliance blocks. In this way, actors can manage ambiguity related to this disruptive change and can keep options open through their multiple partners (Gomes-Casseres, 2001).

Concerning the managerial implications, the network positioning decision involves a short-term and a long-term consideration. Actors should be aware that alliance networks are very dynamic – just like the strategic alliances they are based on – and they should be aware of the fact that being embedded in an alliance network in terms of occupying a certain network position can have an enabling as well as a constraining effect on the actors' innovative performance. For the short term, it might be beneficial for an actor to leave a rigid alliance group to move into new windows of opportunity. This strategic move might improve its innovative performance in the short run as this actor gets access to new resources and capabilities. In the long term however, this move can involve reputation effects as this firm leaves its partners, which can result in the breaking up of alliances. This can destroy the social capital the firm has built up in the past. Thus, escaping from the constraining effects of embeddedness by changing the short-term network position in order to improve innovative output can harm a firm's long-term network position as it has broken up partnerships in the past for short-term benefits. It will be difficult for this actor to revive these severed ties because of the reputation effects that apply in the network.

Concerning our theoretical considerations, we believe that in order to explain the formation of alliance blocks in an inter-organizational network in relation to innovative performance, one has to go beyond the resource-based view (Dyer and Singh, 1998) and incorporate the ability of responsiveness of firms, which is the point of departure in the behavioural and dynamic capabilities theories (Kogut, 1988; Teece and Pisano, 1994). Further research could therefore develop more rich theoretical implications of the firm-level resource-based view (Dyer and Singh, 1998) to the group level in order to better explain the group-versus-group dynamics in the formation of alliance blocks.

9. Summary

This volume has made a conceptual as well as an empirical contribution to the academic work on social structure and its effect on alliance formation patterns in social networks (see for example Gulati, 1995a, 1998; Granovetter, 1992; Gulati and Gargiulo, 1999; Walker et al., 1997). Moreover, this study has contributed conceptually as well as empirically to the current body of literature on social embeddedness and network positioning strategies of firms in alliance networks – in alliance blocks in particular – and their effect on innovative performance (for example Rowley et al., 2000; Gulati et al., 2000; Gargiulo and Benassi, 2000; Coleman, 1988; Burt, 1992) under changing technological conditions (Madhavan et al., 1998; Bower and Christensen, 1995). We have conducted a quantitative research project based on an extensive literature study and on an empirical analysis of the CATI alliance database in the microelectronics industry.

We have empirically studied network evolution in inter-organizational networks in a high-tech industry from a longitudinal perspective. We have further developed the concept of alliance blocks or 'technology driven constellations' (Gomes-Casseres, 1996) by adopting a social network perspective (for example Nohria, 1992; Gulati, 1998). We identified the social mechanisms that follow from embeddedness and from investing in social capital (Gulati, 1995a; Gulati and Gargiulo, 1999). These social mechanisms cause enabling and constraining effects of embeddedness in technology alliance blocks and induce network dynamics. They have an enabling effect as well as a constraining effect in the evolution of alliance blocks and inter-organizational networks, as they cause the network to self-generate and self-transform. The network evolutionary perspective showed us that alliance blocks are socio-technical systems, where inter-organizational networks co-evolve with their technologies and can result in the formation of alliance blocks.

We found descriptive empirical evidence for the enabling effect of embeddedness in technology alliance blocks, as we were able to empirically point at the social mechanisms local search and replication of ties that bring about the formation of alliance blocks. More specifically we found that as an alliance network grows, the likelihood of alliance block formation also increases as firms replicate their ties within the subgroup as they continue to engage in new relationships over time.

Concerning the constraining effect of embeddedness in technology alliance blocks that follows from the social mechanism of similarity that causes attraction and interaction, we found descriptive empirical evidence that through the replication of their existing ties in alliance blocks, alliance block members tend to become more similar over time. This can lead to decreasing learning effects and even cause a state of over-embeddedness. Concerning the constraining role of embeddedness that follows from the social mechanism of relational inertia, we found descriptive evidence for lock-out effects. We found that in a growing network, newcomers form groups among themselves as they are locked out of established groups.

To explore the formation of alliance blocks in relation to innovative performance, we address the strategic behavioural (Kogut, 1988) and dynamic capabilities (Teece and Pisano, 1994) approach to explain the strategic moves of firms in dynamic context. This means, we have placed collective collaborative agreements, like alliance blocks, in the context of competitive rivalry to enhance market power. These multiple strategic alliances are vital instruments for developing dynamic capabilities and improving innovative performance, as these alliances act as vehicles to complement the firm's needs in changing markets and technological environments. To gain more insight into the characteristics of alliance block members, we have empirically studied the specific attributes that distinguish them from non-alliance block members and which can influence their innovative capability. We revealed that alliance block members, compared to their non-block counterparts, are likely to apply for more patents and hold more central positions in the network. Furthermore alliance block members have larger revenues and undertake more R&D-intensive research. In addition, they are also inclined to originate from an Asian background in comparison to non-alliance block members.

We found a positive relation for alliance block membership and innovative performance. This implies that alliance block members are productive in their joint innovative efforts. However as we have seen in our analyses, there is a limit to the advantage of operating in a closed social system such as an alliance block. Over time those alliance block members start to suffer from relational and technological over-embeddedness, caused by relational inertia and the increasing similarity of firms within the alliance blocks.

In this study we also addressed the moderating effect of turbulence in the technological environment on network positioning strategies and innovative performance of technology-based firms. We found that pursuing a block membership strategy under cumulative technological change is valuable for innovative performance. Under these conditions joint innovative efforts through alliance block formation ameliorate the coordination and the linking of these innovations together enhancing and extending the underlying sustaining technology (Tushman and Anderson, 1986). Block members exploit their existing capabilities, as they are linked up with firms in their own technology cluster to improve their innovative performance. This supports our expectation that these specific technological conditions foster the virtues of closure and block membership as they have a positive impact on innovative activity.

A striking finding was that block members do not only perform well under cumulative change, but also under disruptive change. This means that dealing with disruptive change in a group of alliances proves to be an effective strategy. Apparently the trust-based governance we find in alliance blocks is not only beneficial under conditions of cumulative change, as we found, it is also beneficial under turbulent technological changes.

We found that non-block members have a low innovative performance under cumulative technological change as we expected. Another striking finding was that – contrary to our expectations – non-alliance block members do not perform well in a disruptive and turbulent technological environment. We expected that disruptive change would be favourable for non-alliance block members. Under those circumstances, they could flexibly manoeuver into new technological opportunities, without facing the inertia that block members would have to face. Actually our findings indicate that non-block members perform worse than block members under conditions of disruptive technological change.

These findings also have implications regarding our arguments on over-embeddedness, in the sense that our hypotheses suggest that in case of disruptive change the constraining effects of embeddedness will surface earlier for block members, as in this situation rigidity and inertial effects would be effectuated earlier. Again, we could not find evidence for this, because our results indicated that alliance block members perform well under disruptive change.

Appendix I

Table A I Technology profile in microelectronics 1989–91

Microelectronics technology classes 1989–91	257	326	327	345	361	365	438	505
AMD	15	39	21	0	3	24	21	0
IBM	107	52	42	161	81	62	88	8
INTEL	8	17	13	2	4	44	29	0
MOTOROLA	126	46	78	11	71	34	140	4
NEC	136	63	57	14	16	143	55	4

Continued	708	709	710	711	712	713	714
AMD	14	1	12	9	10	1	7
IBM	36	108	97	112	51	56	144
INTEL	9	2	11	26	16	2	7
MOTOROLA	30	4	13	15	12	17	32
NEC	47	15	27	40	43	8	41

257 Active solid-state devices
326 Electronic digital logic circuitry
327 Electricity battery or capacitor
345 Computer graphic processing, operator interface processing
361 Electricity: electrical systems and devices
365 Static information storage and retrieval
438 Semiconductor device manufacturing
505 Superconductor technology
708 Electrical computers: arithmetic processing and calculating
710 Electrical computers: input/output
711 Electrical computers: memory
712 Electrical computers and digital processing systems
713 Electrical computers and digital processing systems: support
714 Error detection/correction and fault detection/recovery

Appendix II

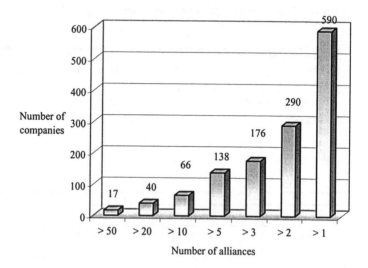

Figure A II Determining the cut-off point for the sample

COMPANIES IN SAMPLE

Minnesota Mining & Mfg. Co.
Acer Corp.
Acorn Computer Group Plc
Advanced Risc Machines Holdings Ltd (ARM)
Aerospatiale (SNIAS)
Alcatel N.V.
Advanced Micro Devices Inc.
American Microsystems Inc. (AMI)
Austria Mikro Systeme Int. AG (AMS)
Analog Devices Inc ADI
Apple Computer Inc.
Asea A.B.
Askey Computer Corp
American Telephone & Telegraph Co.
AT&T Bell Laboratories
Boeing Computer Services
British Aerospace Plc.
Brown Boveri & Co.A.G. (BBC)
Bull Groupe S.A.
Cadence Design Systems Inc
Cal Technologies Cabinets Inc
Canon Inc.
Control Data Corp .(CDC)
Centric Engineering Systems
Chipcom Corporation
Chips & Technologies
Cirrus Logic
Cisco Systems Inc
Compaq Computer Corp.
Convex Computer
Costa Mesa
Compañia Telefónica Naçional de E.
Cypress Semiconductor

DB Networks Ltd
Digital Equipment Corp .(DEC)
Ericsson A.B., Telefon
Fairchild Semiconductor Corp.
Fiat SpA.
Force Computers Inc.
Fujitsu Ltd.
General Electric Co .(GE)
GEC General Electric Co PLC
GEC Plessey Semiconductors Inc GPS
Gemplus
General Instrument Corp.
Goldstar Co.
Harris Corp.
Hitachi Ltd.
Honeywell Inc.
Hewlett-Packard Co.
Huawei Technology Co Ltd (Shenzhen)
Hughes Aircraft Co.
Hyundai Corp.
Hyundai MicroElectronics
Int. Business Machines Corp. (IBM)
Int. Computers Ltd. (ICL)
IMEC (Belgium)
Infineon Technologies AG
Inmos
Intel Corp.
Intelligent Aerodynamics
Int. Rectifier Corp.
Italtel Spa
Irvine Sensors
Eastman Kodak Co.
Lockheed Missile and Space Research
Lear Siegler Inc.
LSI Logic
Lucent Technologies Inc

Matra-Harris Semiconducteurs (MHS)
Matsushita Elect.Industrial Co.Ltd.
Micron Technology Inc.
Microsoft Corp.
Mietec Alcatel
Mips Computer Systems
Mitsubishi Corp
Mitsubishi Electric Corp
Mitsui Group
Mostek
Motorola Inc.
Nasa
National Semiconductor Corp.
National Cash Register Corp. (NCR)
Nippon Electric Corp.(NEC)
NEC Technologies Inc
NMB Semiconductor
Nippon Telegraph & Telephone (NTT)
Oki Electric Industry Co.
Olivetti SpA.
Orbit
Philips Gloeilampenfabrieken N.V.
Philips Semiconductors
Plessey Co.
Prime Computer Inc.
Ramtron International Corp
Renssellaer Polytechnic Institute
Rice University
Ricoh Co.
Rohm GmbH.

Samsung Electronics America Inc
Samsung Electronics Co
Samsung Co.Ltd.
Sandia National Laboratories
Sandisk
Sanyo Electric Co.Ltd.
Seagate Technology Inc.
Sequoia Systems
Saint-Gobain S.A.
SGS-Ates
SGS-Thomson Microelectronics NV
SGS-Thomson Microelectronics
Sharp Corp.
Siemens A.G.
Signetics
Silicon Graphics
Sony Corp.
STMicroelectronics
Standard Microsystems Corp.
Sumitomo Corp
Sun Microsystems
Sun Moon Star Co. Ltd
Synoptics communication Inc
Tecom Corp. Ltd
Telfin
Thomson S.A.
Texas Instruments Inc.
Toshiba Corp.
Toyota Motor Corp.
Unisys Corp.
Videologic
VLSI Technology
Westinghouse Electric Corp
Yokogawa-Hokushin Elect.Works
(YEW)
Zilog Inc
Zyxel Communication Corp

Appendix III

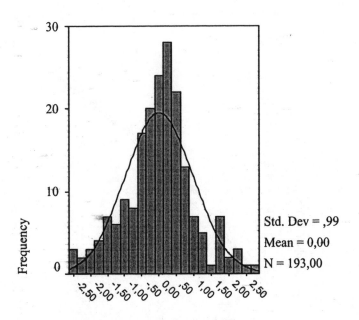

Figure A III Histogram of normal distribution

Appendix IV

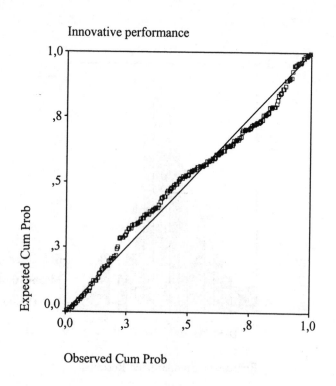

Figure A IV Normal pp plot of residual

Appendix V

Table A V Derivation independent variable

Company	Period	Block membership	Years in block
3M	1980–82 *(t1)*	0	0
3M	1981–83 *(t2)*	0	0
3M	1982–84 *(t3)*	0	0
3M	1983–84 *(t4)*	0	0
3M	1984–86 *(t5)*	1	1
3M	1985–87 *(t6)*	1	2
3M	1986–88 *(t7)*	1	3
3M	1987–89 *(t8)*	0	0
3M	1988–90 *(t9)*	0	0
3M	1989–91 *(t10)*	0	0
3M	1990–92 *(t11)*	0	0
3M	1991–93 *(t12)*	0	0
3M	1992–94 *(t13)*	0	0
3M	1993–95 *(t14)*	0	0
3M	1994–96 *(t15)*	0	0
3M	1995–97 *(t16)*	0	0

Appendix VI

TESTING FOR MULTICOLLINEARITY FOR MODELS 1, 2 AND 3

In order to detect possible multicollinearity in our data we performed a number of multicollinearity tests. We address tolerance (1 – Ri-square), where Ri-square is the multiple correlation coefficient when the ith independent variable is predicted from the other independent variables. Here a small tolerance indicates multicollinearity. A common cut-off threshold is a tolerance value of 0.10. Values near 0.10 indicate multicollinearity. As we can see from the tables in this appendix, no variables come even close to the threshold level of tolerance (0.1). Our analysis shows that for models 1, 2 and 3 all tolerances are above 0.821; this indicates a low level of multicollinearity.

Subsequently, we use *VIF*, which is a reciprocal of tolerance. It measures the inflation in variances of the parameter estimates due to multicollinearities. High *VIF* values, corresponding to VIF values above 10, indicate multicollinearity. The largest *VIF* level for both Model 1, 2 and 3 is 1.218, which thus does not show any signs of multicollinearity.

In order to validate the findings we calculated some additional diagnostic measures of multicollinearity. Thereto, we calculated the amount of multicollinearity present with the condition index. The condition index represents the amount of multicollinearity of combinations of variables in the data set. Condition indices are used to flag excessive collinearity in the data. The most commonly used threshold value for high level of multicollinearity is 30. That means that the existence of a condition index greater than 30 could indicate a high level of multicollinearity among variables. In our analysis, there are no values over the threshold limit. Our highest condition index for models 1, 2 and 3 is 6.039. Again, no multicollinearity is detected.

Additionally, we checked the variance proportion, which represents the proportion of variance for each regression coefficient (and its associated variable) attributable to each condition index. Coefficients that have high proportions for the same condition index represent that the associated variables are largely responsible for the amounts of multicollinearity identified by the condition index. A value of 0.5 is commonly used as a cut-off point. However there only exists a multicollinearity problem if the condition index is greater than 30 and accounts for a substantial proportion of variance (at least 0.5 and preferably 0.9 or above) for two or more coefficients. In our analysis we find some variance proportions with values between 0.5 and 0.9, however, the associated condition indexes are not higher than 6.039, which means that again no multicollinearity is detected.

Table A VI.1 Multicollinearity statistics

	Model 1				Model 2				Model 3			
	Tol.	VIF	Eigenv.	CI	Tol.	VIF	Eigenv.	CI	Tol.	VIF	Eigenv.	CI
Constant			3.251	1			3.033	1			2.952	1.000
Block membership	0.925	1.081	1.037	1.770								
Years in block					0.883	1.133	0.125	4.918				
Europe dummy	0.897	1.115	0.819	1.992	0.909	1.101	1.033	1.714	0.907	1.102	1.083	1.651
Asia dummy	0.870	1.149	0.419	2.785	0.892	1.122	0.852	1.887	0.883	1.132	0.858	1.855
R&D intensity	0.883	1.132	8.915E-02	6.039	0.821	1.217	0.397	2.766	0.821	1.218	0.395	2.734
Spec.	0.842	1.188	0.384	2.908	0.859	1.164	0.560	2.328	0.860	1.163	0.587	2.243
Interaction effect squared									0.868	1.152	0.126	4.845

Table A VI.2 Variance proportions

		Constant	1	1a	1b	2	3	4	5
Model 1	Constant	0.01	0.03			0.01	0.01	0.03	0.01
	1.Block membership	0.00	0.01			0.27	0.40	0.03	0.00
	2. Europe	0.00	0.00			0.49	0.17	0.16	0.01
	3. Asia	0.03	0.05			0.05	0.25	0.78	0.06
	4. R&D ratio	0.94	0.08			0.17	0.06	0.01	0.82
	5. Spec.	0.02	0.83			0.00	0.11	0.00	0.09
Model 2	Constant	0.02		0.03		0.02	0.01	0.04	0.02
	1a. Years in Block	0.92		0.01		0.18	0.19	0.00	0.76
	2. Europe	0.00		0.03		0.17	0.49	0.02	0.00
	3. Asia	0.01		0.11		0.57	0.07	0.07	0.00
	4. R&D Ratio	0.03		0.11		0.03	0.11	0.85	0.11
	5. Spec.	0.02		0.71		0.04	0.13	0.03	0.10
Model 3	Constant	0.02			0.03	0.02	0.01	0.04	0.02
	1b. Interaction Effect [2]	0.92			0.01	0.18	0.20	0.00	0.77
	2. Europe	0.01			0.10	0.09	0.44	0.02	0.00
	3. Asia	0.00			0.11	0.65	0.01	0.05	0.00
	4. R&D Ratio	0.04			0.10	0.03	0.09	0.85	0.12
	5. Spec.	0.01			0.65	0.03	0.24	0.04	0.08

References

Achilladelis, B., A. Schwartzkopf and M. Cines (1987), 'A study of innovation in the pesticide industry: analysis of the innovation record of an industrial sector', *Research Policy*, **16** (2–4), 175–212.

Acs, Z. and D. Audretsch (1989), 'Patents as a measure of innovative activity', *Kyklos*, **42** (2), 171–80.

Ahmadjian, C.L. and J.R. Lincoln (2001), 'Keiretsu, governance and learning: case studies in change from the Japanese automotive industry', *Organization Science*, **12** (6), 683–701.

Ahuja, G. and R. Katila (2001), 'Technological acquisitions and the innovation performance of acquiring firms: a longitudinal study', *Strategic Management Journal*, **22** (3), 197–220.

Alba, R.D. (1973), 'A graph-theoretic definition of a sociometric clique', *Journal of Mathematical Sociology*, **3**, 113–26.

Archibugi, D. (1992), 'Patenting as an indicator of technological innovation: a review', *Science and Public Policy*, **6**, 357–58.

Aspden, H. (1983), 'Patent statistics as a measure of technological vitality', *World Patent Information*, **5**, 170–73.

Basberg, B. (1987), 'Patents and the measurement of technological change: a survey of the literature', *Research Policy*, **16** (2–4), 131–41.

Borgatti, S.P., M.G. Everett and L.C. Freeman (1999), *UCINET 5 for Windows: Software for Social Network Analysis,* Natick: Analytic Technologies.

Borgatti, S.P., M.G. Everett and L.C. Freeman (2002), *UCINET 6 for Windows: Software for Social Network Analysis*, Harvard: Analytic Technologies.

Borgatti, S.P., M.G. Everett and P.R. Shirey (1990), 'LS Sets, lambda sets, and other cohesive subsets', *Social Networks* **12**, 337–58.

Bourdieu, P. (1986), 'The Forms of Capital', in J.G. Richardson (ed.), *Handbook of Theory and Research for the Sociology of Education*, New York: Greenwood, 241–58.

Bourdieu, P. and L. Wacquant (eds) (1992), *An Invitation to Reflexive Sociology*, Chicago, IL: University of Chicago Press.

Bovasso, G. (1996), 'A network analysis of social contagion processes in an organizational intervention', *Human Relations*, **49** (11), 1419–36.

Bower, J.L. and C.M. Christensen (1995), 'Disruptive technologies: catching the wave', *Harvard Business Review*, **73** (1), 43–54.

Brass, D.J., K.D. Butterfield and B.C. Skaggs (1998), 'Relationships and unethical behavior: a social network perspective', *Academy of Management Review*, **23** (1), 14–31.

Burt, R.S. (1992), 'The social structure of competition', in N. Nohria and Robert G. Eccles (eds), *Networks and Organizations: Structure, Form and Action*, Boston, MA: Harvard Business School Press, pp. 57–91.

Burt, R.S. (2000), 'Pre-print for chapter', in R.I. Sutton and B.M. Staw (eds), *Research in Organizational Behavior*, **22**, Greenwich, CT: JAI Press.

Chesbrough, H.W. and D.J. Teece (1996), 'When is virtual virtuous? Organizing for innovation', *Harvard Business Review*, **94** (1), 65–73.

Chung, S.A., H. Singh and K.Lee (2000), 'Complementarity, status similarity and social capital as drivers of alliance formation', *Strategic Management Journal*, **21**, 1–20.

Cohen, W. and R. Levin (1989), 'Empirical studies of innovation and market structure', in R. Schmalensee and R.D. Willig (eds), *Handbook of Industrial Organization*, **II**, Amsterdam: Elsevier, 1059–107.

Cohen, W.M. and D.A. Levinthal (1990), 'Absorptive capacity: a new perspective on learning and innovation', *Administrative Science Quarterly*, **35**, 128–52.

Coleman, J.S. (1988), 'Social capital in the creation of human capital', *American Journal of Sociology*, **94**, 95–120.

Coleman, J.S. (ed.) (1990), *Foundations of Social Theory*, Cambridge, MA: Harvard University Press.

Comanor, W. and F. Scherer (1969), 'Patent statistics as a measure of technological change', *Journal of Political Change*, **77** (3), 392–98.

Cyert, R. and J.G. March (eds) (1963), *A Behavioral Theory of the Firm*, Englewood Cliffs, NJ: Prentice-Hall.

Das, T.K. and B. Teng (2002), 'Alliance constellations: a social exchange perspective', *Academy of Management Review*, **27** (3), 445–56.

Das, T.K. and B. Teng (2003), 'Partner analysis and alliance performance', *Scandinavian Journal of Management*, **19**, 279–308.

De Man, A. and G.M. Duysters (2003), *Collaboration and Innovation: A Review of the Effects of Mergers, Acquisitions and Alliances on Innovation*, Report for Dutch Ministry of Economic Affairs.

Dore, R. (1983), 'Goodwill and the spirit of market capitalism', *British Journal of Sociology*, **34**, 459–82.

Dosi, G. (1988), 'Sources, Procedures and Microeconomic Effects of Innovation', *Journal of Economic Literature*, 26: 1120–71.

Doz, Y.L. and G. Hamel (eds) (1998), *Alliance Advantage, The Art of Creating Value Through Partnering*, Boston, MA: Harvard Business School Press.

Duysters, G.M. (ed.) (1996), *The Dynamics of Technical Innovation: the Evolution and Development of Information Technology*, Cheltenham, UK and Brookfield, US: Edward Elgar Publishing.

Duysters, G.M. and A. De Man (2003), 'Transitory alliances: an instrument for surviving turbulent industries?' *R&D Management*, **33** (1), 49–58.

Duysters, G.M., R. Guidice, B. Sadowski and A. Vasudevan (2001), 'Inter alliance rivalry, theory and application to the global telecommunications industry', in M. Hughes (ed.), *International Business and its European Dimensions*, London: MacMillan.

Duysters, G.M. and J. Hagedoorn (1993), 'The cooperative agreements and technology indicators (CATI) information system', unpublished working paper, MERIT, University of Maastricht.

Duysters, G.M. and J. Hagedoorn (1995), 'Strategic groups and inter-firm networks in international high-tech industries', *Journal of Management Studies*, **32**, 361–81.

Duysters, G.M. and J. Hagedoorn (1998), 'Technological convergence in the IT industry: the role of strategic technology alliances and technological competencies', *International Journal of the Economics of Business*, **5**, 355–68.

Duysters, G.M. and J. Hagedoorn (2000), 'Core competences and company performance in the world-wide computer industry', *Journal of High Technology Management Research*, **11** (1), 75–91.

Duysters, G.M. and J. Hagedoorn (2001), 'Do company strategies and structures converge in global markets? Evidence from the computer industry', *Journal of International Business Studies*, **32** (2), 347–56.

Duysters, G.M., J. Hagedoorn and C.E.A.V. Lemmens (2003), 'The effect of alliance block membership on innovative performance,

Revue d'Economie Industrielle, Special Issue 2nd and 3rd trimester, **103**, 59–70.

Duysters, G.M. and C.E.A.V. Lemmens (2003), 'Alliance group formation: enabling and constraining effects of embeddedness and social capital in strategic technology alliance networks', *International Studies of Management and Organization*, **33** (2), 49–68.

Dyer, J.H. and H. Singh (1998), 'The relational view: cooperative strategy and sources of inter-organizational competitive advantage, *Academy of Management Review*, **23** (4), 660–79.

Ellis, C.J. and D. Fausten (2002), 'Strategic FDI and industrial ownership structure', *Canadian Journal of Economics*, **35** (3), 476–94.

Everett, M.G. and S.P. Borgatti (1999a), 'Models of core/periphery structures', *Social Networks*, **21**, 375–95.

Everett, M.G. and S.P. Borgatti (1999b), 'Peripheries of cohesive subsets', *Social Networks*, **21**, 397–407.

Fershtman, M. (1997), 'Cohesive group detection in a social network by the segregation matrix index', *Social Networks*, **19**, 193–207.

Foster, R.N. (ed.) (1986), *Innovation: The Attacker's Advantage*, New York: Summit Books.

Freeman, L.C. (1979), 'Centrality in social networks', *Social Networks*, **1**, 215–39.

Friedkin, N.E. (1984), 'Structural cohesion and equivalence explanations of social homogeneity, *Sociological Methods and Research*, **12**, 235–61.

Garcia-Pont, C. and N. Nohria (2002), 'Local versus global mimetism: the dynamics of alliance formation in the automobile industry, *Strategic Management Journal*, **23** (4), 307–21.

Gargiulo, M. and M. Benassi (2000), 'Trapped in your own net? Network cohesion, structural holes and the adaptation of social capital', *Organization Science*, **11** (2), 183–96.

Gnyawali, D.R. and R. Madhavan (2001), 'Cooperative networks and competitive dynamics: a structural embeddedness perspective', *Academy of Management Review*, **26** (3), 431–45.

Gomes-Casseres, B. (1994), 'Group versus group: how alliance networks compete', *Harvard Business Review*, **72** (4), 62–70.

Gomes-Casseres, B. (ed.) (1996), *The Alliance Revolution: The New Shape of Business Rivalry*, Cambridge, MA: Harvard University Press.

Gomes-Casseres, B. (2001), 'The logic of alliance fads: why collective competition spreads', in M.P. Koza and A.Y. Levin (eds), *Strategic Alliances and Firm Adaptation: A Coevolution Perspective*, M.E. Sharpe.

Gomes-Casseres, B., A.B. Jaffe and J. Hagedoorn (2002), 'Knowledge flows across firm and national boundaries', unpublished working paper.

Gómez, C., B.L. Kirkman and D.L. Shapiro (2000), 'The impact of collectivism and in-group/out-group membership on the evaluation generosity of team members', *Academy of Management Journal*, **43** (6), 1097–106.

Granovetter, M. (1973), 'The strength of weak ties', *American Journal of Sociology*, **78**, 1360–80.

Granovetter, M. (1985), 'Economic action and social structure: the problem of embeddedness', *American Journal of Sociology*, **91** (3), 481–510.

Granovetter, M. (1992), 'Problems of explanation in economic sociology', in N. Nohria and Robert G. Eccles (eds), *Networks and Organizations: Structure, Form and Action*, Boston, MA: Harvard Business School Press, 25–56.

Griliches, Z. (1990), 'Patent statistics as economic indicators: a survey, *Journal of Economic Literature*, **28** (4), 1661–797.

Guidice, R.M., A. Vasudevan and G.M. Duysters (2003), 'From "me against you" to "us against them": alliance formation based on inter-alliance rivalry', *Scandinavian Journal of Management*, **19**, 135–52.

Gulati, R. (1995a), 'Social structure and alliance formation patterns: a longitudinal analysis, *Administrative Science Quarterly*, **40**, 619–52.

Gulati, R. (1995b), 'Does familiarity breed trust? The implications of repeated ties for contractual choice in alliances', *Academy of Management Journal*, **38** (1), 85–112.

Gulati, R. (1998), 'Alliances and networks', *Strategic Management Journal*, **19**, 293–317.

Gulati, R. and M. Gargiulo (1999), 'Where do inter-organizational networks come from?' *American Journal of Sociology*, March, **104**, 1439–93.

Gulati, R., N. Nohria and A. Zaheer (2000), 'Strategic networks', *Strategic Management Journal*, **21**, 203–15.

Hagedoorn, J. (1993), 'Understanding the rationale of strategic technology partnering: interorganizational modes of cooperation and sectoral differences', *Strategic Management Journal*, **14**, 371–85.

Hagedoorn, J. (1996), 'Trends and patterns in strategic technology partnering since the early seventies', *Review of Industrial Organization*, **11**, 606–16.

Hagedoorn, J. and G.M. Duysters (2002), 'Learning in dynamic inter-firm networks – the efficacy of multiple contacts', *Organization Studies*, **23**, 525–48.

Hagedoorn, J. and J. Schakenraad (1992), 'Leading companies and networks of strategic alliances in information technologies', *Research Policy*, **21**, 163–90.

Hagedoorn, J. and J. Schakenraad (1994), 'The effect of strategic technology alliances on company performance', *Strategic Management Journal*, **15**, 291–309.

Hamel, G. (1991), 'Competition for competence and inter-partner learning within international strategic alliances', *Strategic Management Journal*, **12**, 133–39.

Hannan, M.T. and J. Freeman (1977), 'The population ecology of organizations', *American Journal of Sociology*, **82**, 929–64.

Hannan, M.T. and J. Freeman (1984), 'Structural inertia and organizational change', *American Sociological Review*, **49**, 149–64.

Henderson, R. and W. Mitchell (1997), 'The interactions of organizational and competitive influences on strategy and performance', *Strategic Management Journal*, **18**, 5–14.

Hobday, M. (1997), 'The technological competence of European semiconductor producers', *International Journal of Technology Management*, **14** (2, 3, 4), 401–14.

Holbrook, D., W.M. Cohen, D.A. Hounshell and S. Klepper (2000), 'The nature, sources, and consequences of firm differences in the early history of the semiconductor industry', *Strategic Management Journal*, **21**, 1017–41.

Jelinek, M and C.B. Schoonhoven (eds) (1990), *The Innovation Marathon: Lessons from High Technology Firms*, Oxford: Basil Blackwell.

Kash, D.E. and R.W. Rycoft (2000), 'Patterns of innovating complex technologies: a framework for adaptive network strategies', *Research Policy*, **29**, 819–31.

Knoke, D. and J.H. Kuklinski (eds) (1982), *Network Analysis, Series: Quantitative Applications in the Social Sciences*, Beverly Hills and London: Sage Publications.

132   *Innovation in Technology Alliance Networks*

Kogut, B. (1988), 'Joint ventures: theoretical and empirical perspectives', *Strategic Management Journal*, **9** (4), 319–32.

Kogut, B. (1989), 'The stability of joint ventures: reciprocity and competitive rivalry', *Journal of Industrial Economics*, **38**, 183–93.

Kogut, B. and U. Zander (1993), 'Knowledge of the firm, combinative capabilities, and the replication of technology', *Organization Science*, **3** (3), 383–97.

Koka, B.R. and J.E. Prescott (2002), 'Strategic alliances as social capital: a multidimensional view', *Strategic Management Journal*, **23**, 795–816.

Krackhardt, D. (1992), 'The strength of strong ties: the importance of philos in organizations', in N. Nohria and Robert G. Eccles (eds), *Networks and Organizations: Structure, Form and Action*, Boston, MA: Harvard Business School Press, 216–39.

Lane, P.J. and M. Lubatkin (1998), 'Relative absorptive capacity and interorganizational learning', *Strategic Management Journal*, **19**, 461–77.

Langlois, R.N. and W.E. Steinmüller (2000), 'Strategy and circumstance: the response of American firms to Japanese competition in semiconductors, 1980–1995', *Strategic Management Journal*, **21**, 1163–73.

Larson, A. (1992), 'Network dyads in entrepreneurial settings: a study of the governance of exchange processes', *Administrative Science Quarterly*, **37**, 76–104.

Leonard-Barton, D. (ed.) (1995), *Wellsprings of Knowledge*, Cambridge, MA: Harvard Business School Press.

Levin, R.C., A.K. Klevorick, R.R. Nelson, S.G. Winter, R. Gilbert and Z. Griliches (1987), 'Appropriating the returns from industrial research and development: comments and discussion', *Brookings Papers on Economic Activity*, **3**, 783–831.

Levinthal, D.A. and M. Finchmann (1988), 'Dynamics of interorganizational attachments: auditor–client relationships', *Administrative Science Quarterly*, **5**, 583–601.

Levitt, B. and J.G. March (1988), 'Organizational learning', *Annual Review of Sociology*, **14**, 319–40.

Lin, N. (1999), 'Building a network theory of social capital', *Connections*, **22** (1), 28–51.

Lippman, S.A. and R.P. Rumelt (1982), 'Uncertain imitability: an analysis of interfirm differences in efficiency under competition', *Bell Journal of Economics*, **13** (2), 418–38.

Madhavan, R., B.R. Koka and J.E. Prescott (1998), 'Networks in transition: how industry events (re)shape interfirm relationships', *Strategic Management Journal*, **19**, 439–59.

Mowery, D.C. (ed.) (1988), *International collaborative Ventures in US Manufacturing*, Cambridge: Ballinger.

Mowery, D.C., J.E. Oxley and B.S. Silverman (1996), 'Strategic alliances and interfirm knowledge transfer', *Strategic Management Journal*, **17**, 77–91.

Mueller, D.C. and J.E. Tilton (1969), 'Research and development costs as a barrier to entry', *Canadian Journal of Economics*, **2**, 570–79.

Mytelka, L.K. (ed.) (1991), 'Strategic partnerships and the world economy', London: Pinter Publishers.

Nahapiet, J. and S. Ghosal (1998), 'Social capital, intellectual capital, and the organizational advantage', *Academy of Management Review*, **23** (2), 242–66.

Nelson, R.R. and S.G. Winter (eds) (1982), *An Evolutionary Theory of Economic Change*, Cambridge, MA: Belknap Press.

Nohria, N. (1992), 'Is a network perspective a useful way of studying organizations?' in N. Nohria and Robert G. Eccles (eds), *Networks and Organizations: Structure, Form and Action*, Boston, MA: Harvard Business School Press, 1–22.

Nohria, N. and C. Garcia-Pont (1991), 'Global strategic linkages and industry structure', *Strategic Management Journal*, **12**, 105–24.

Nonaka, I. (1994), 'A dynamic theory of organizational knowledge creation', *Organization Science,* **5** (1), 14–24.

Osborn, R.N. and J. Hagedoorn (1997), 'The institutionalization and evolutionary dynamics of inter-organizational alliances and networks', *Academy of Management Journal*, **40** (2), 261–79.

Park, S.H., R.R. Chen and S. Gallagher (2002), 'Firm resources as moderators of the relationship between market growth and strategic alliances in semiconductor start-ups', *Academy of Management Journal*, **45** (3), 527–45.

Park, S.H. and Y. Luo (2001), 'Guanxi and organizational dynamics: organizational networking in Chinese firms', *Strategic Management Journal*, **22**, 455–77.

Patel, P. and K. Pavitt (1991), 'Large firms in the production of the world's technology: an important case of 'non-globalization'', *Journal of International Business Studies*, **22**, 1–21.

Pfeffer, J. and P. Nowak (1976), 'Joint ventures and interorganizational interdependence', *Administrative Science Quarterly*, **21**, 398–418.

Pfeffer, J. and G. Salancik (eds) (1978), *The External Control of Organizations: a Resource Dependence Perspective*, New York: Harper & Row.

Podolny, J.M. and T.E. Stuart (1995), 'A role-based ecology of technological change', *American Journal of Sociology*, **5**, 1224–60.

Podolny, J.M., T.E. Stuart and M.T. Hannan, (1996), 'Networks, knowledge, and niches: competition in the worldwide semiconductor industry, 1984–1991', *American Journal of Sociology*, **102** (3), 659–89.

Powell, W.W. (1990), 'Neither market nor hierarchy: network forms of organization', *Research in Organizational Behavior*, **12**, 295–336.

Powell, W.W. and P. Brantley (1992), 'Competitive cooperation in biotechnology: learning through networks', in N. Nohria and Robert G. Eccles (eds), *Networks and Organizations: Structure, Form and Action*, Boston, MA: Harvard Business School Press, 366–94.

Powell, W.W., K.W. Koput and L. Smith-Doer (1996), 'Interorganizational collaboration and the locus of innovation: networks of learning in biotechnology', *Administrative Science Quarterly*, **41**, 116–45.

Prahalad, C.K and G. Hamel (1990), 'The core competence and the corporation', *Harvard Business Review*, **68** (3), 71–91.

Putnam, R.D. (ed.) (1993), *Making Democracy Work: Civic Traditions in Modern Italy*, Princeton, NJ: Princeton University Press.

Raub, W. and J. Weesie (1990), ' Reputation and efficiency in social interactions: an example of network effects', *American Journal of Sociology*, **96**, 626–54.

Richardson, G.B. (1972), 'The organization of industry', *Economic Journal*, **82**, 509–18.

Rosenkopf, L. and A. Nerkar (2001), 'Beyond local search: boundary-spanning, exploration, and impact in the optical disk industry', *Strategic Management Journal*, **22**, 287–306.

Rowley, T., D. Behrens and D. Krackhardt (2000), 'Redundant governance structures: an analysis of structural and relational embeddedness in the steel and semiconductor industries', *Strategic Management Journal*, **21**, 369–86.

Sakakibara, M. (2002), 'Formation of R&D consortia: industry and company effects', *Strategic Management Journal*, **23**, 1033–55.

Sarkar, M.P., R. Echambadi and J.S. Harrison (2001), 'Alliance entrepreneurship and firm market performance', *Strategic Management Journal*, **22**, 701–11.

Scherer, F. (1965), 'Corporate inventive output, profitability and sales growth', *Journal of Political Economy*, **73** (3), 290–97.

Sherwin, C. and R. Isenson (1967), 'Project hindsight', *Science*, **156**, 1571–77.

Silverman, B.S. and J.A.C. Baum (2002), 'Alliance-based competitive dynamics', *Academy of Management Journal*, **45** (4), 791–806.

Steensma, H.K. and K.G. Corley (2000), 'On the performance of technology-sourcing partnerships: the interaction between partner interdependence and technology attributes', *Academy of Management Journal*, **43** (6), 1045–67.

Stoelhorst, J.W. (2002), 'Transition strategies for managing technological discontinuities: lessons from the history of the semiconductor industry', *International Journal of Technology Management*, **23** (4), 261–86.

Stuart, T.E. and J.M. Podolny (1996), 'Local search and the evolution of technological capabilities', *Strategic Management Journal*, **17**, 21–38.

Stuart, T.E. and J.M. Podolny (2000), 'Positional causes and consequences of alliance formation in the semiconductor industry', in J. Weesie and W. Raub (eds), *The Management of Durable Relations: Theoretical Models and Empirical Studies of Households and Organizations*, Amsterdam: Thelathesis.

Tabachnick, B.G. and L.S. Fidell (eds) (1996), *Using Multivariate Statistics*, 3rd edn, New York: Harper Collins College Publishers.

Teece, D. and G. Pisano (1994), 'The dynamic capabilities of firms: an introduction', *Industrial and Corporate Change*, **3** (3), 537–56.

Thomas, H. and N. Venkatraman (1988), 'Research in strategic groups: progress and prognosis', *Journal of Management Studies*, **25** (6), 537–55.

Trajtenberg, M. (1990), *Economic Analysis of Product Innovation: the Case of Ct Scanners*, Cambridge, MA: Harvard University Press.

Tsai, W. (2000), 'Social capital, strategic relatedness and the formation of intra-organizational linkages', *Strategic Management Journal*, **21**, 925–39.

Tsai, W. (2001), 'Knowledge transfer in intraorganizational networks: effects of business position and absorptive capacity on business unit innovation and performance', *Academy of Management Journal*, **44** (5), 996–1004.

Tushman, M.L. and P. Anderson (1986), 'Technological discontinuities and organizational environments', *Administrative Science Quarterly*, **31**, 439–65.

Utterback, J.M. (1974), 'Innovation in industry and the diffusion of technology', *Science*, **198**, 620–26.

Uzzi, B. (1997), 'The social structure and competition in interfirm networks: the paradox of embeddedness', *Administrative Science Quarterly*, **42**, 35–67.

Vanhaverbeke, W., G.M. Duysters and N.G. Noorderhaven (2002), 'External technology sourcing through alliances or acquisitions: an analysis of the application-specific integrated circuits industry', *Organization Science*, **13** (6), 714–33.

Vanhaverbeke, W. and N.G. Noorderhaven (2001), 'Competition between alliance blocks: the case of the RISC microprocessor technology', *Organization Studies*, **22** (1), 1–30.

Walker, G., B. Kogut and W. Shan (1997), 'Social capital, structural holes and the formation of an industry network', *Organization Science*, **8** (2), 109–25.

Wasserman, S. and K. Faust (eds) (1994), *Social Network Analysis, Methods and Applications*, Cambridge: Cambridge University Press.

West, J. (2002), 'Limits to globalization: organizational homogeneity and diversity in the semiconductor industry', *Industrial and Corporate Change*, **11** (1), 159–88.

Index

Absorptive capacity
  pre-alliance technological
    overlap 44, 47–48
Alliance block formation
  technological drivers 7
  social drivers 8
Alliance block formation
  microelectronics industry 7
  motives 34, 42
  network evolutionary
    perspective 7
  social network perspective
    4, 7
  socio-technical system 9
Alliance block measures
  lambda sets 15, 23
  line connectivity 23–26
  subgroup cohesion 24
  in-group/out-group ratio 38
Alliance block members
  attributes 56
  innovative performance 57,
    66–67, 73, 81
Alliance block membership
  11, 22, 65, 72, 85
Alliance blocks
  characteristics 15, 23, 35
  theoretical perspectives 52,
    54
  cliques 15
  constellations 5–6
  technology alliance blocks
    5, 8, 15

technology-driven groups,
    5, 15
Alliance network formation
  dynamics 32
  motives 3, 42
  social network perspective
    4, 7
  social structural context 4
AMD (Advanced Micro
    Devices) 25, 47

Brokerage 63, 65–66

Cliques 15
Closure 63, 65–66
Competition
  group vs. group 5–6
  group-based 6
  technology competition 6
Constellations 5-6
  See Alliance blocks
Control Variables 27, 72–73,
    88
Co-opetition 5
  See Competition
Cumulative Change 82, 85–87

Data sources 16–17, 37, 46,
    70, 84
  MERIT-CATI database 16
Discriminant analysis 56–59
Disruptive change 83, 85–87
Dynamic capabilities 53–54

Embeddedness
  See Social embeddedness
Enabling embeddedness
  local search 34
  replication of ties 34, 44
Endogenous constraints
  lock-in / lock-out effects,
  43, 45, 49, 68
  strategic gridlock 45–51,
  68
Endogenous dynamics 3
Evolutionary theory 8
Exogenous dynamics 3

Home region 27

IBM 47
Incremental change
  See Cumulative change
Innovative performance 17,
  21–22, 67–69, 81–84
Intel 47
Interaction effect 84–85
Inverted U-shape 69, 71, 76
  curvilinear 69, 71, 76

Lambda sets 15, 23
Local search 34
Lock-in / lock-out 43, 45, 49,
  68

Methodology 17–20, 37–38,
  47–48, 70–71, 84–85
  moving window 19
  time lag 19
Microelectronics industry 27–
  31
  history 27
  characteristics 28–29
Mostek 47
Motorola 47

National background
  See Home region
NEC 47
Networks
  See Social networks

Ordinary least square
  regression (OLS) 70, 84
Organizational learning theory
  8
Over-embeddedness 45, 48,
  68–69
  relational inertia 45–51, 68
  strategic gridlock 45–51,
  68
  technological similarity 36,
  44, 48

Preferential relations 8, 34
Positional Embeddedness 64

R&D intensity 27
Relational embeddedness
  65
Relational inertia 45–51, 68
Replication of ties 34, 44
  motives 35–37
Reputation effects 36
Resource dependency 3, 53

Sample 37, 46, 70, 84
Siemens 47
Similarity 36, 44, 48
  similarity breeds attraction,
  44
  interaction breeds similarity
  44
Social capital 32–33
  group-level 33
Social contagion 43
Social embeddedness 4, 64–
  65

performance 63–65, 66–69, 80–81
debate 11, 63
constraining effect 42–46
enabling effect 32–37
relational, positional and structural 64–65
Social mechanisms, constraining 43
Social network perspective 4
Social networks 32
Socio-technical system 9
Strategic behavior approach 53–54
Strategic gridlock 45–51, 68
Strategic groups 54–56
Strategic technology alliances 16, 37
examples 16
Strong ties 34
Structural embeddedness 64–65
Structural inertia theory 8
Structure-loosening events 12, 83–84
Structure-reinforcing events 82
Subgroup cohesion 24
Sunk costs 36
Switching costs 36

Technological change 83
Technological regime 81–82
Technology positioning 11
Technology profile 47, 115
Technological specialization 27, 72, 88
Transaction cost perspective 35

Unethical behavior 36

Variables
control variables, 27, 72, 88
dependent variable, 21, 71, 85
independent variable, 22, 72, 85
moderator variable 85–88

Zilog 47
Z-score normalization 87–88

.